# Microbial Pathogens and Human Diseases

# Microbial Pathogens and Human Diseases

**Naveed Ahmed Khan**
School of Biological & Chemical Sciences
Birkbeck College
University of London
London
UK

Science Publishers
Enfield (NH) Jersey Plymouth

*CIP data will be provided on request.*

Science Publishers
234 May Street
Post Office Box 699
Enfield, New Hampshire 03748
United States of America

*www.scipub.net*

General enquiries : *info@scipub.net*
Editorial enquiries : *editor@scipub.net*
Sales enquiries : *sales@scipub.net*

Published by Science Publishers, Enfield, NH, USA
An imprint of Edenbridge Ltd., British Channel Islands
Printed in India



ISBN 978-1-57808-535-4

# Introduction

Infectious diseases due to microbial pathogens have been causing human misery and death, since the emergence of the human species. The most distressing aspect is that every time we find a way to overcome them by developing preventative measures (vaccines) or chemotherapeutic artillery (antimicrobials) and claim to be the dominant species in the world, pathogens develop means (evolve) to resist the treatment approaches available at our disposal. With increased global travel, mass production and transport of food, crowded urban populations and continuing world conflicts, we have transformed this planet to the benefit of pathogens. What we must recognise is that any species on this planet has an obvious goal, to get fitter and rule, after all, its' survival of the fittest. Only if we can learn from the pathogens and put our differences aside and unite against our common enemy, that has caused immense human suffering, in the form of plagues, black death etc. With the present complacency towards microbial pathogens, it is difficult to see how we can remain to be the dominant species on this planet and it's only a matter of time that we are wiped out of existence. This book combines basic sciences, clinical medicine and pathology, with an attempt to help basic scientists and clinicians to collaborate in presenting information about medical microbiology and host-parasite relationships. This book guides you through the basics of microbiology, immunology, and infectious diseases and helps you understand the significance of host-parasite relationships in the development or prevention of infection. Overall, this book should be of interest both to students (Universities and Medical schools) as well as experts in the relevant fields or those who want to pursue research in this interesting and important area.

# Contents

CHAPTER

1

# Microbial or Infectious Diseases: Introduction

## 1. DISEASE

A disease is any abnormal condition of the body that causes discomfort, anxiety, dysfunction, or distress to the person affected. This term is used broadly to include injuries, disabilities, symptoms, syndromes (set of symptoms), unusual behaviour, and abnormal structures and functions. More specifically, this term is used to describe an atypical condition in the living organism that interferes with the normal bodily function of the organism resulting in symptoms (characteristics observed or felt by the patient), signs (observed or measured by others), and ill-health. Disease is also referred to as morbidity. Of interest, the term 'infection' is used to describe when a pathogen invades a host, while 'disease' is described when the invading pathogen alters normal body function. There are different kinds of human diseases (Table 1). Among them, diseases caused by agents such as viruses, bacteria, protozoa, fungi and metazoa (usually helminths) are called 'infectious diseases'. Except metazoa, other pathogens are very small (invisible to naked eye) and are called microbial pathogens and their diseases are referred to as 'microbial infectious diseases' or simply 'microbial diseases' (Fig. 1).

## 2. MICROBIAL INFECTIOUS DISEASES

These include diseases caused by agents that can be transmitted from person to person via direct contact or indirectly via a contaminated object. For example, the common cold, or HIV (human immunodeficiency virus), a causative agent of AIDS (Acquired Immuno Deficiency Syndrome), or

**Table 1** Types of diseases.

| *Types of Diseases* | *Cause* |
|---|---|
| **Infectious diseases** | Microbial agents, transmitted to susceptible hosts via exposure to contaminated environment or infected organisms. |
| **Inherited diseases** | Abnormal genes, passed from one generation to the next. |
| **Neoplastic diseases** | Abnormal cell growth leading to formation of benign or malignant tumours. Causes include genetic, environmental factors, chemicals, radiation and viruses. |
| **Immunity-related diseases** | These develops when immune system fails so body is unable to defend itself or becomes abnormal so immune system begins attacking normal tissues, e.g., autoimmune diseases. |
| **Degenerative diseases** | Associated with aging. For example, there are significant reductions in cardiac efficiency, kidney filtration, etc. |
| **Nutritional deficiency diseases** | Caused by lack of appropriate nutrition, vitamins, proteins, carbohydrates, etc. |
| **Endocrine diseases** | Abnormal production of hormones. |
| **Latrogenic / Nosocomial diseases** | Hospital acquired, results from activity or treatment of physicians, e.g., post-surgery. |
| **Environmental diseases** | Results from the exposure to environmental poisons. |
| **Idiopathic diseases** | Causes are not yet known. |

bacterial meningitis are all examples of microbial or infectious diseases. In contrast, diseases such as diabetes and cancer are non-infectious diseases. The majority of microbes associated with the human body (such as the gastrointestinal tract) are part of the normal flora. However, less than one percent of them have the ability to cause harm to human tissues resulting in pathology (i.e., tissue abnormality) that is referred to as 'disease'. An organism that has the ability to produce disease is called a pathogenic organism or simply a pathogen. If the organisms are very small and cannot be observed by the naked eye, they are called microbial pathogens (also called infectious agents) and these can produce diverse types of human infections.

## 3. WHY STUDY INFECTIOUS DISEASES–GLOBAL PERSPECTIVE

Infectious diseases have been causing human misery and death since the emergence of the human species. Enlarged spleens most likely due to malaria were found in Egyptian mummies, more than 5,000 years ago. Malaria antigens were found in the skin and lung samples of mummies from 3204 BC. Over the last 5,000 years, we have made significant advances

| | | |
|---|---|---|
| **Primary infections** | : | In these infections, microbes are the primary cause of infection and exhibit apparent clinical symptoms. |
| **Secondary infections** | : | Microbial invasion subsequent to primary infection. |
| **Mixed infections** | : | Two or more microbes infecting the same tissue. |
| **Acute infections** | : | Disease progresses rapidly, within hours or days. |
| **Chronic infections** | : | Disease progresses slowly, takes months or years. |
| **Sub-clinical infections** | : | No detectable clinical symptoms. |
| **Dormant infections** | : | Microbes uses host as a carrier (carrier state). |
| **Accidental infections** | : | Environmental or accidental exposure to microbes. |
| **Opportunistic infections** | : | These are caused by microbes when host defences are compromised. |
| **Localized infections** | : | Infections are limited to a small area or to a specific tissue. |
| **Generalized infections** | : | These are disseminated to many tissues or different body regions. Examples include Gram negative bacteremia. |
| **Pyogenic infections** | : | These infections involve pus-formation and can be caused by *Staphylococcal* and *Streptococcal* infections. |
| **Retrograde infections** | : | Bacteria ascend in a duct or tube against the flow of secretions. Examples include *E. coli* urinary tract infections. |
| **Fulminant infections** | : | These infections occur suddenly and intensely. Examples include airborne *Yersinia pestis* (pneumonic plague). |

**Fig. 1** Types of microbial infections.

in medical sciences and claim to be the dominant species in the world. Yet, malaria alone is killing more than a million people per year, worldwide, while HIV/AIDS is causing nearly 2.7 million annual deaths, and tuberculosis contributes to 1.7 million deaths. Our available artillery, i.e., antimicrobials/vaccines has helped us increase the chances of the survival of our species against the fittest and perfectly placed species, i.e., pathogens. However, clear evidence is emerging that some pathogens are continuously developing means to resist the treatment approaches available at our disposal. The process of drug resistance in pathogens is occurring at a much faster rate than our ability to produce new drugs or new approaches to interfere with the disease process. It is only a matter of time before the arsenal of antimicrobials available at our disposal becomes obsolete. The result will be a repeat of plague (black death), cholera, influenza-like outbreaks wiping out whole communities and even nations. During the 14th and 15th centuries, Europe's population was halved with the outbreaks of smallpox and plague. In the 1800s, outbreaks of puerperal sepsis (streptococcal infection) caused more than 70% deaths in new mothers in Europe. The outbreak of influenza in 1918 spread throughout the world causing millions of deaths (around 30 million) destroying whole communities and having devastating effects on the economies. And even today, despite the discovery of antibiotics/drugs and the available supportive care, infectious diseases have remained the leading causes of human deaths. Annually, there are approximately 14 million deaths due to infectious diseases, worldwide, more than half of them occurring in children. This is caused by just the top ten infectious diseases (Table 2). Apart from the fatal consequences, hundreds of millions of people are left disabled or orphaned due to infectious diseases. In addition to the direct role of pathogens in causing human misery and loss of life, it is now well-established that infectious diseases also contribute to non-infectious diseases. For example, chronic infections due to human papillomavirus results in the development of cervical cancer, the most common cancer in the developing countries. Similarly, chronic infections due to hepatitis B and hepatitis C can result in liver cancer, while schistosomiasis can result in bladder cancer. Overall, there is clear need to study infectious agents, understand their biology and mechanisms of diseases in an attempt to identify targets for preventative and/or therapeutic approaches.

**Table 2** Leading infectious causes of death in 2002 alone (source: World Health Organization, 2002).

| *Cause* | *Rank* | *Estimated No. of Deaths* |
|---|---|---|
| Acute lower respiratory infections | 1 | 3,884,000 |
| HIV/AIDS | 2 | 2,777,000 |
| Diarhoeal diseases | 3 | 1,798,000 |
| Tuberculosis | 4 | 1,556,000 |
| Malaria | 5 | 1,272,000 |
| Measles | 6 | 611,000 |
| Pertussis | 7 | 294,000 |
| Tetanus | 8 | 214,000 |
| Sexually transmitted diseases (STD, excluding HIV) | 9 | 180,000 |
| Meningitis | 10 | 173,000 |

## 4. THE FORGOTTEN WORLD

The infectious diseases largely occur in developing countries. The most distressing aspect is that the majority of these infections can be prevented simply by employing basic health measures and the burden can be significantly reduced (more than 70%) by employing improved sanitary measures and patient access to hospitals and/or basic treatment. A simplified view of this problem is the lack of available sources, lack of appropriate health systems and/or drugs and overcrowding. For example, there are nearly one and a half billion people living in the developing countries with an income of less than one US dollar per day. There are one in three children who are malnourished and one in five children who are not fully immunized. These conditions provide perfect opportunities for pathogens to emerge and reemerge. During the last 50 years, the neglect of the wealthy nations has pushed the developing countries deeper into social and economical problems resulting in their continued under-development. For example, malaria alone costs Africa billions of dollars every year. The usual targets of infectious diseases are children or the common people. The loss of children has devastating effects on communities, while the loss of the head of household has catastrophic effects on families both in terms of social consequences and the loss of earnings and access to costly health care costs. It is not surprising that infectious diseases are closely linked with poverty and keep communities in a constant cycle of hardship and ill-health with the notion "poverty causes illness and illness causes poverty". For example, HIV/AIDS alone has left eight million children orphaned. However, infectious diseases no longer remain a problem of the developing countries or countries with limited resources. For example, tuberculosis and diphtheria once thought to be under control, are reemerging and spreading throughout the world, especially in the developed countries. Today, tuberculosis is killing approximately 1.7 million people per year, while eight million people become newly infected. Similarly, the recent 1996 outbreak of polio in Greece, Albania and Yugoslavia had worrisome effects on the health professionals of this (i.e., polio) long forgotten disease.

## 5. WHY CAN'T WE GET RID OF INFECTIOUS DISEASES?

The first major breakthrough in the control of infectious diseases came with the pioneering work of Edward Jenner. In 1796, he introduced a cure against smallpox and proved that injecting a cowpox virus into humans can prevent them from the lethal attack of the smallpox virus. His observations were so novel that they have led to the discovery of vaccination. To date, smallpox is the first and only infectious disease that has been truly eradicated from the world. Later, the use of chemicals in the treatment of infectious diseases was first established by Paul Ehrlich in early 1900s and his investigations led to the treatment of diphtheria. Later, the discovery of antibiotics in 1928 was made (the first one discovered was penicillin by Alexander Fleming) and their role as therapeutic measures against infectious diseases was shown by Howard Florey and Ernst Chain in 1935, and their potential was fully explored following the Second World War. By 1960s, several physicians claimed that "the threat of infectious diseases with serious consequences exists no more". Antibiotics were claimed to be magic bullets or miracle drugs. Later, antifungal, antiparasitic and more recently antiviral compounds were identified and they are collectively known as antimicrobials. Indeed, antimicrobials, together with improved sanitation and appropriate health care systems have largely reduced the risks of infectious diseases in developed nations, but the situation in developing countries has remained bleak as ever. Now, several decades' later, infectious diseases have continued to cause more than 14 million deaths annually, both in the developing and developed countries and at the same time, the few remaining available antimicrobials are becoming less effective in the treatment of infectious diseases. There are several reasons for the continued threat posed by infectious agents (Table 3).

**Table 3** Factors contributing to the emergence, spread and reemergence of infectious diseases.

| | |
|---|---|
| **Environmental changes** | These may be due to agriculture, dams, de/re-forestation, flood/drought, climate changes and may contribute to emergence, spread or reemergence of infectious diseases. |
| **Human activities** | Population growth, immigration, use of high population densities such as day care centres, army, prisons, war or civil conflicts, sexual behaviours or use of intravenous drugs. |
| **Industry** | Globalization of food supplies, changes in food processing, transplantation, immunosuppressive drugs, widespread use of antimicrobials. |
| **Emergence of a resistant strain** | Bacterial evolution in response to a given environment, i.e., antibiotics or harsh environmental conditions. |
| **Public health measures** | Lack or reduction in preventative measures, i.e., appropriate sanitary measures, vector control measures, vaccination. Other factors include poor nutrition and water supplies, lack of personal hygiene, limited access to hospitals/treatment. |

1. The world is shrinking and the means of travelling (such as air travelling) allow pathogens (or pathogen-infected individuals) to move from endemic areas to new susceptible populations with relative ease, i.e., within hours.
2. The world population is rapidly increasing, from 2 billion in 1930, to an estimate of 9 billion in 2050 (Table 4). This is despite the fact that the available resources have remained the same. This has resulted in the densely populated areas especially inner cities with poor sanitation and poverty which provide a foothold for pathogens to emerge and reemerge.
3. Mass population movements.
4. Natural disasters such as flooding as well as environmental changes due to global warming contribute to sustaining and the spreading of infectious diseases.
5. Man-created disasters such as wars help ensure pathogen survival, emergence and/or reemergence.
6. More worryingly, is the emergence of antimicrobial-resistant pathogens which are contributing to increased costs of care, increased and prolonged morbidity and increased mortality and this has raised alarms over the limited number of effective antimicrobials available at our disposal.
7. With the increasing costs in the development of antimicrobials (approx. US $500 million for each compound), it is not surprising that we have seen only a limited success in the identification of new antimicrobial compounds during the last few decades.

**Table 4** Increase in the world's population.

| *Year* | *Estimated world population* |
|---|---|
| 1930 | 2 billion |
| 1976 | 4 billion |
| 1992 | 5.5 billion |
| 2050 | 9 billion – expected |

## 6. ANTIMICROBIAL RESISTANCE: A CLEAR AND PRESENT DANGER

A few years after the discovery of penicillin, the emergence of penicillin-resistant strains of *Staphylococcus aureus* was observed. Within the next few years, antimicrobial resistance was observed among dysentery-causing *Shigella* spp. and *Salmonella* spp. and now it has become a serious concern for public health. For example, in 1990, almost all cholera isolates in New Delhi (India) were sensitive to furazolidone, ampicillin, co-trimoxazole and nalidixic acid. In 2000, these drugs became largely obsolete in the treatment of cholera. Similarly, multi drug-resistant tuberculosis has spread throughout the developing and developed countries and is no longer confined to immunocompromised or HIV patients, while resistant malaria is killing millions of people annually. Today, 98% of all gonorrhoea cases in South East Asia are multi-drug resistant, while 60% of all visceral leishmaniasis cases in India and 60% of hospital-acquired infections in the developed nations are now caused by drug-resistant microbes. For example, vancomycin-resistant Enterococcus (VRE) and methicillin-resistant *Staphylococcus aureus* (MRSA) have recently emerged as major threats to public health with economical and social implications. In fact, the only available antimicrobial left to treat MRSA is vancomycin and now recently it has emerged that vancomycin-intermediate *Staphylococcus aureus* (VISA) is exhibiting some levels of resistance against vancomycin. Pathogens develop drug resistance by a process known as natural selection, in which a sub-population of pathogens evolves to develop drug resistance. Drug resistant genes are then passed to the next generation by replication or to other related microbes by conjugation or by plasmids, carrying the drug-resistant genes which move from one organism to another. Over-prescription of antibiotics by physicians, incomplete courses of antimicrobials by patients and counterfeit or inappropriate drugs expedite this process and contribute to the emergence of drug-resistant pathogens.

In addition to the clinical complications with fatal consequences, the most obvious problem of antimicrobial resistance is that many of the cheap but valuable available drugs are no longer of any use. For example, the costs of drugs used against tuberculosis were US $20, while the costs of drugs against multi-drug resistant tuberculosis are around US $2,000. And there are only a handful of antimicrobials left, which can be used in the treatment of drug-resistant microbes. It appears that despite our advances in medical sciences, pathogens remain one step ahead, or find ways to adapt or cope with any challenges we throw at them (Table 5).

## 7. PORTALS OF ENTRY

Given the access and/or the opportunity, pathogenic microbes can attack nearly all tissues/organs in the human body. Some produce infections at

**Table 5** The history of medicine.

- 2000 B. C. – Here, eat this root.
- 1000 A.D. – That root is heathen. Here, say this prayer.
- 1850 A.D. – That prayer is superstition. Here, drink this potion.
- 1920 A.D. – That potion is snake oil. Here, swallow this pill.
- 1945 A.D. – That pill is ineffective. Here, take this penicillin.
- 1955 A.D. – 'Oops'....bugs mutated. Here, take this tetracycline.
- 1960-1999 – 39 more 'oops'...Here, take this more powerful antibiotic.
- 2000 A.D. – The bugs have won! Here, eat this root.

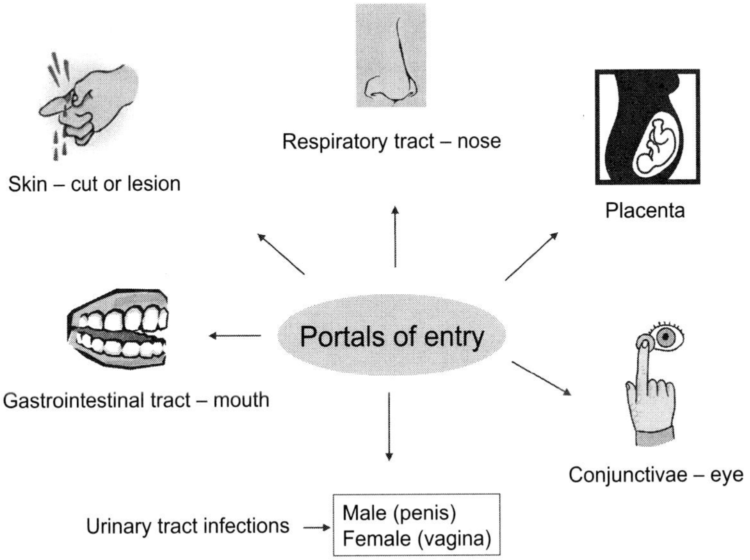


**Fig. 2** The portals of entry of infectious agents.

their portal of entry and may disseminate to other organs (Fig. 2) to produce multiple infections, while others only can cause tissue/organ-specific infections. Below are examples of major portals of entry for microbial pathogens. It is noteworthy that many microbes may reside in their host as part of the normal flora and produce infections under specific conditions such as a weaker immune system. There are three major portals of entry as described below.

**7.1. Skin – scratch or injury**
**7.2. Mucosal membranes**
**7.2.1. Respiratory tract – nose**
**7.2.2. Conjunctivae – Eye**
**7.2.3. Gastrointestinal tract – mouth**
**7.2.4. Urinary tract infections**
**7.2.4.1. Urethra**
**7.2.4.2. Vagina**
**7.3. Placenta**

## 7.1 Skin – scratch or Injury

The skin is composed of packed, dead skin cells that normally act as a barrier to invading pathogens. The presence of a cut or injury to the skin can provide a route to the pathogens to access the human body. Some pathogens have the ability to digest the outer layer of the skin by secreting toxins allowing their entry into the human body.

### 7.2 Mucosal Membranes

These membranes line the body cavities that are open to the environment and provide a moist, warm environment that is rich in nutrients and is hospitable to pathogens. Although these membranes are protected by strong host defences, pathogens have the ability to evade the defences and produce infections. For example, the respiratory tract is protected by strong air movements, complex anatomical structures and ciliary epithelium. In contrast, the stomach has acidic pH and other chemical disinfectants that are highly destructive to microorganisms; however some pathogens can survive this onslaught to produce infection. It is not surprising that the respiratory tract is the most common portal of entry for pathogens.

### 7.3 Placenta

Pathogens can transmit from the infected mother to the embryo or the foetus via the placenta route and may result in abortion, severe birth defects, death or other complications.

## 8. PATHOGEN RESERVOIR

The majority of microbial pathogens cannot survive for a long period of time outside of their host and use animals, humans or non-livings as reservoirs. In animal reservoir, pathogens use animals as their usual hosts but spread to humans to produce disease. Such diseases are called zoonotic diseases and are transmitted through direct contact with animals or their waste. However, these pathogens do not usually transfer back to animals. Other reservoirs include humans, which may also act as a reservoir for human pathogens. For example, asymptomatic-infected individuals, who transmit the disease to the susceptible hosts, as is the case for AIDS or syphillis. In addition, the pathogen reservoir may include soil, water, food that may be contaminated via faeces or urine.

## 9. MODES OF TRANSMISSION

Infectious diseases are transmitted to humans via exposure to a contaminated environment or infected individuals/animals (Fig. 3).

### 9.1 Contact Transmission

This may involve direct contact with the infected host such as handshaking, sex, kissing, bite or indirect contact via drinking from the same glass, sharing

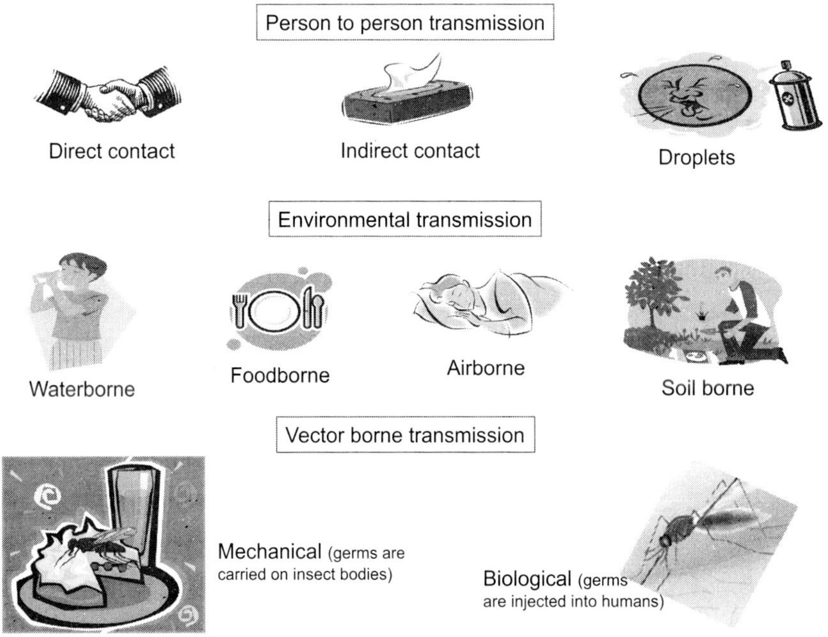


**Fig. 3** The modes of transmission of infectious diseases.

of an inert object (such as towel or toothbrush), or droplets from sneezing or coughing.

### 9.2 Vehicle Transmission

This may be airborne (dust particles), waterborne (streams or swimming pools) or foodborne (e.g., uncooked meat).

### 9.3 Vector Transmission

This may be mechanical, i.e., insect acts as a carrier and transmits the pathogen to a susceptible host or biological, i.e., pathogen multiplies within the insect before transmission to the susceptible host.

## 10. THE WAY FORWARD

It has become patently obvious that the continued complacency by the wealthy nations towards the developing countries has made everyone vulnerable to the emerging and reemerging infectious diseases such as tuberculosis. It must be understood that infectious diseases are a common target that continue to seriously threaten the human species. Infectious agents from all over the world have a common objective, i.e., to target other living organisms to ensure the survival of their species. We must learn from them and unite in confronting pathogens, have early warning systems in place to prevent the spread of infections, rapidly increase the arsenal of drugs for our urgent needs and develop alternative strategies for therapeutic interventions. The urgent needs are as follows:

1. Seriously ill children may be infected with more than one pathogen and the diagnosis is difficult. Thus combined therapy (which may include oral rehydration salts to treat diarrhoea, antibiotics to treat pneumonia, antimalarial drugs, vitamins and mineral supplements) should prove highly effective.
2. Increased awareness and better feeding practices will significantly reduce the risk of infectious diseases.
3. Early diagnosis is of crucial value for the successful treatment of infectious diseases.
4. Identification of the risk factors should enhance our ability to interfere with the infectious diseases. For example, malaria is spread by mosquitoes and the use of bed-nets while sleeping should help reduce the number of malaria-associated deaths.
5. Availability of drugs: As indicated above, millions of people in developing countries are dying needlessly from diseases that could be easily treated with inexpensive drugs. Access of these drugs to the needy can help develop these communities.
6. Education of safe sex, sexually transmitted diseases, hygiene and their associated risks.
7. Mass immunizations have proven effective in eradicating smallpox and similar approaches for other infectious diseases should be the objective for future.
8. Strengthen health services and delivery systems in developing countries.
9. Expansion of surveillance systems to alert unexpected outbreaks, the emergence of new diseases and increased drug resistance.
10. Need to develop novel drugs as well as to slow the rate of drug resistance.
11. Development of diagnostic tools, new drugs and vaccines that can further improve our ability to target infectious diseases.

CHAPTER

# 2

# Major Human Microbial Infectious Diseases

## 1. The study of microbial infectious diseases: a physician's and a microbiologist's perspective

Microbial diseases are studied depending on the professional expertise. A physician handles patients and classifies diseases based on its site of infection. From a physician's perspective, infectious diseases can be classified as

- Skin infections
- Respiratory infections
- Gastrointestinal infections
- Urinary tract infections
- Soft tissue infections
- Central nervous system infections
- Bone and joint infections
- Cardiovascular infections
- Disseminated infections

This scheme helps localizing the infection so that any test for the correct diagnosis of the disease can be carried out on the infected part of human body. In addition, this approach helps in the application of treatment on the infected site or identifies the best available approaches for the therapeutic interventions. In this chapter, I have used this scheme to describe major human microbial infectious diseases.

However, the microbiologist's perspective is to look at the infectious diseases from the microbial side. Usually their interest lies in the classification of the infectious diseases based on the pathogens which are as follows:

- Viruses
- Bacteria
- Fungi
- Protozoa

Of course, it's the combination of both that result in the correct diagnosis of the disease, leading to the identification and application of the best available treatment measures which result in the favourable outcome of the disease for the patient.

## 2. Human body

The basis of the human body is atoms. One or several atoms form molecules, which perform different functions in the human body or form various types of cells. There are more than 200 different types of cells in the human body. Cells contain several molecules and are the smallest units of life. Depending on the type, cells may function independently, e.g., macrophages police the body to eliminate pathogens. Other cells group together (or work together) and form tissues. There are four types of tissues which are as follows:

1. Epithelial tissue: these cover body surfaces exposed to the environment, body cavities, etc., e.g., the gastrointestinal tract is covered with epitheliaum.
2. Connective tissue: these fill spaces and provides structural support, e.g., bones, fat deposits.
3. Muscle tissue: these provides bodily movement by contraction.
4. Nervous tissue: these conduct information from one part of body to another in the form of electrical impulses.

Several tissues work together to perform specific tasks and are called 'organs', e.g., the heart is made of epithelial, connective, muscular and nervous tissues to pump blood through the body and this organ system is called 'cardiovascular system'. There are 11 organs systems which work together to make a human body. These organ systems are described below:

(1) Integumentary (i.e., covering) system – this includes the skin, hair, nail and exocrine glands that protect the body, store fat, excrete water and salt in the form of sweat.

(2) Skeletal system – this includes bones and cartilages which provide structural support for the body. It stores minerals such as calcium, phosphate, fats and produces red blood cells and other blood elements.

(3) Muscular system – muscles attached to the skeleton (skeletal muscles) provides bodily movements and support, while other muscle types provide organ support, e.g., cardiac muscle.

(4) Nervous system – the primary function of the nervous system is to monitor internal and external conditions using sensory receptors, coordinate the sensory information and direct a response using other systems. It is further divided into

i. Central nervous system (brain and spinal cord), the brain maintains the memory, intelligence, emotion. The CNS is responsible for processing sensory information and co-ordinating a response.
ii. Peripheral nervous system provides a communication link between the CNS and the rest of the body. It monitors internal and external conditions and reports it to the CNS and then carries messages from the CNS to the body.

(5) Endocrine system consists of endocrine cells that secrete hormones which affect the activities of other cells (Fig. 1).

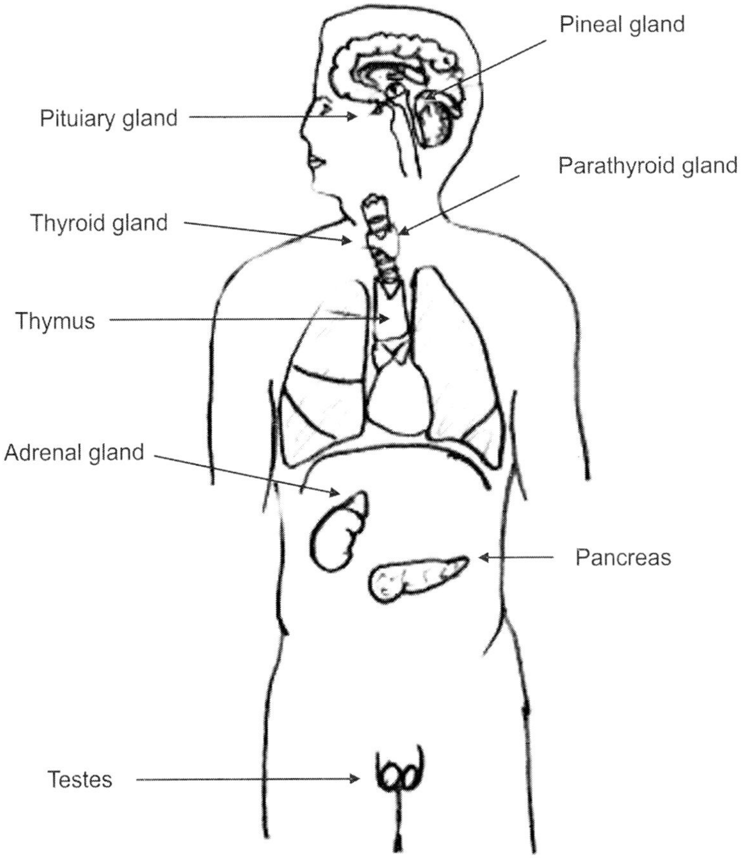


**Fig. 1(A)** Endocrine system.

| Gland | Hormone | Function |
|---|---|---|
| Pituitary gland | Thyroid-stimulating hormone | Stimulate thyroid hormone secretion |
| | Adrenocorticotropic hormone | Stimulate glucocorticoid secretion |
| | Follicle-stimulating hormone | Promote egg development in females and sperm production in males |
| | Luteinizing hormone | Stimulate ovulation (egg release) in females and testosterone in males |
| | Prolactin | Production of milk |
| | Growth hormone | Stimulate cell division and growth |
| | Antidiuretic hormone | Reduce water loss at kidneys |
| | Oxytocin | Stimulate contraction of uterus smooth muscles in females and prostate gland contraction in males |
| Thyroid gland | Thyroxine | Stimulate rate of body metabolism |
| | Calcitonin | Reduce calcium ions in blood |
| Parathyroid gland | Parathyroid hormone | Increase calcium ions in blood |
| Thymus | Thymosins | Stimulate lymphocyte (white blood cell) development |
| Adrenal gland | Adrenaline (epinephrine) | Stimulate use of glucose and glycogen, increase heart rate |
| | Noradrenaline (norepinephrine) | Similar to adrenaline |
| Pancreas | Insulin | Reduce glucose levels in blood |
| | Glucagon | Increase glucose levels in blood |
| Testes | Testosterone | Stimulate production of sperm and other male characters |
| Ovaries | Estrogen | Stimulate egg development and other female characters |
| | Progesterone | Stimulate embryonic development |
| Pineal gland | Melatonin | Delays sexual maturation |

**Fig. 1(B)** Hormones and their function.

(6) Cardiovascular system consists of heart and blood vessels which extend to the rest of the body. The blood vessels can be divided into two networks, pulmonary circulation where the blood completes a round-trip from the heart to the lungs to exchange carbon dioxide for oxygen and systemic circulation where blood completes a round-trip from the heart to the remaining body to deliver oxygen and nutrients to all tissues.

(7) Lymphatic system is composed of lymphs (fluid similar to blood plasma but contains less proteins) which travel in lymphatic vessels throughout the body and are filtered at specific sites called lymph nodes or lymph glands. Lymphatic system contains lymphocytes (white blood cells) including T-, B- and Natural Killer cells that are produced in the bone marrow. T-lymphocytes migrate to thymus and mature, 'Natural Killer cells' and **B**-lymphocytes (from the **b**one marrow) are released directly into the bloodstream. Lymphocytes play a crucial role in the defence against the pathogens. Other lymphoid

tissues are contained in the spleen, whose function is to screen blood (similar to lymph nodes which screen lymph). The spleen removes abnormal red blood cells, stores iron from red blood cells and monitors and responds to the pathogens in the blood.

(8) Respiratory system consists of the nose, sinuses, pharynx (throat), larynx (voice box), trachea (windpipe), lungs containing bronchi and bronchioles which import oxygen into the body and export carbon dioxide out of the body.

(9) Digestive system consists of the digestive tract (starting with the mouth and ending with the anus as well as the liver, pancreas and salivary glands that mutually digest food, absorbs small organic molecules and excrete waste products.

(10) Urinary tract system consists of kidneys and urethra with the primary function of excretion of waste products from the blood by urine.

(11) Reproductive system is required to produce and transport male and female gametes for reproduction.

## 3. SKIN INFECTIONS

The skin is made of epithelial cells, a layer of tightly joined cells that cover both external and internal surfaces of the body (Fig. 2). The primary function of epithelial cells is to protect internal tissues, prevent dehydration (e.g., the skin protects and resists loss of water), or microbial entry into the human body.

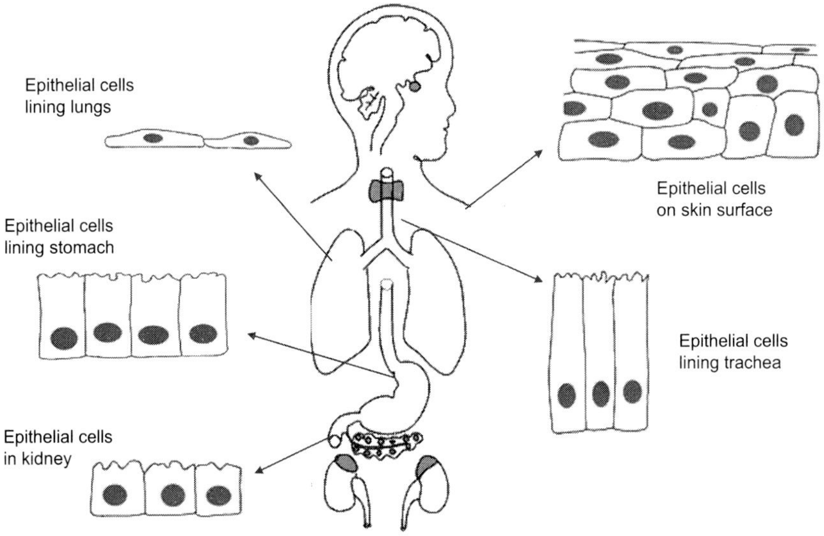


**Fig. 2** Various types of epithelial cells.

## 3.1 Skin Infections – Viral

### *3.1.1 Cutaneous warts*

These are lumps (warts) appearing on the skin, usually limited to children since adults develop immunity against the virus (Fig. 3). There are no symptoms but warts appear as soft, flesh-coloured swellings that can appear anywhere on the body and vary in numbers and size. Treatment is generally not required in children as they gradually develop immunity. In adults, cryotherapy is used. This is performed by freezing the wart with a cryoprobe or with a small swab dipped in liquid nitrogen. Other therapies include the use of salicylate and lactic acid ointment onto the wart. The causative agent is papillomavirus that generally lives in the skin or mucous membranes but occasionally produce visible warts. The virus is transmitted through direct contact with the infected person or rarely through swimming in the same pool as the infected person.

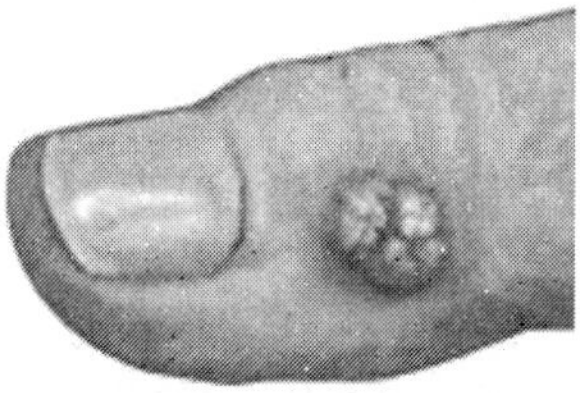

**Fig. 3** Cutaneous wart (source: www.nlm.nih.gov/medlineplus).
*Colour image of this figure appears in the Colour Plate Section at the end of the Book.*

### *3.1.2 Chickenpox (varicella-zoster)*

It is a skin infection (during childhood) with the appearance of a large numbers of itchy, fluid-filled blisters that burst and form crusts. The symptoms of the disease are fever, headache and loss of appetite. The disease is self-limiting but antivirals acyclovir and antihistamines are effective. The causative agent is varicella-zoster virus, which also causes shingles in adults and is transmited via contact with infected individuals.

### *3.1.3 Shingles (also called herpes zoster)*

It is a painful rash with a burning sensation and may also be associated with fever, malaise, headache and red skin (Fig. 4). After the patient suffers from chickenpox, the virus becomes dormant and upon its reemergence (following several years), it causes shingles within 2 – 3 days exhibiting red patches of skin with small vesicles (usually on the trunk, i.e., the upper part of body but rarely involves the face, eyes, mouth and genitals). Within a few weeks, blisters begin to dry and eventually heal. If infection is limited to the skin, treatment may not be necessary but is important if the eye, face, mouth or genitals are involved. The antiviral agent, acyclovir is used to

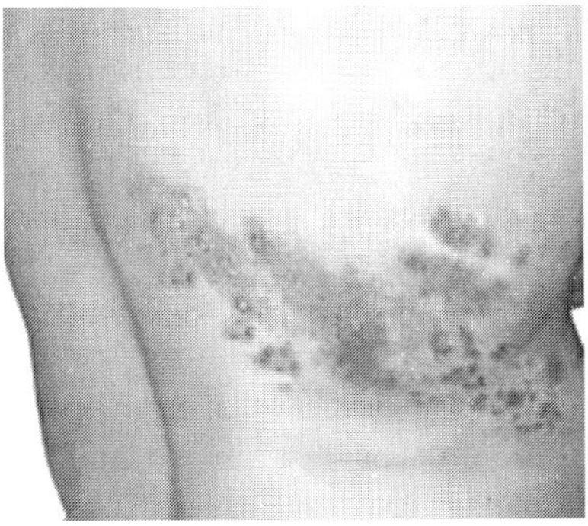

**Fig. 4** Shingles (herpes zoster) (source: www.nlm.nih.gov/medlineplus).
*Colour image of this figure appears in the Colour Plate Section at the end of the Book.*

treat this infection. Other drugs include famciclovir or penciclovir that can be used effectively. The causative agent is varicella zoster virus that also causes chickenpox.

### 3.1.4 *Infectious mononucleosis*

Infection may be associated with fever, generalized lymphadenopathy, sore throat, malaise, tiredness, splenomegaly, hepatomegaly, abnormal liver function tests and atypical lymphocytosis in the peripheral blood. The symptoms appear within several weeks of contracting the virus. The disease is self limiting but antiviral agents such as acyclovir can be used to treat this infection. The causative agent is Epstein barr virus.

### 3.1.5 *Head, foot and mouth disease*

This is a highly infectious disease that can occur at any age but is more common among young children. This is not the foot and mouth disease, which affects cows, sheep and pigs. It is caused by coxsackie A viruses, and characterized by rash on the palms and soles of infected individuals as well as sore areas in the mouth. The rash is normally red and may blister (painful swelling on the skin containing fluid). Other symptoms include fever, sore throat and malaise that develop within five days after contracting virus and patients get better within a few days, without any treatment. The causative agent is coxsackie A virus, which is spread through contact with bodily fluids and by respiratory droplets released into the air by coughing and sneezing.

### 3.1.6. *Cutaneous poxvirus infections*

Poxviruses are divided into three groups:

(i) Orthopoxviruses causing cowpox, monkeypox and smallpox.
(ii) Parapoxviruses causing orf.
(iii) Unclassified viruses causing molluscum contagiosum.

**3.1.6.ia. Cowpox** Cowpox is a zoonotic disease (transmitted from animals) that is usually limited to the skin but may involve the eyes and lymph nodes. It is characterized initially (within a few days) by the appearance of macule (flat small skin spots) at the site of contact with the infected animal. Within two weeks, macules become papules (small solid bumps) and then vesicles (blisters that contain fluid) and finally pustules (raised area of inflamed skin filled with pus) within a month which become hard, are covered in black and eventually heal. The disease is often associated with malaise, fever, lethargy (tiredness), vomiting, and a sore throat lasting about a week. The disease is transmitted through contact with infected cows (person-to-person transmission has not been reported). It is a self-limiting disease and is not treated in the majority of cases but antiviral, cidofovir is a possible choice.

**3.1.6.ib. Monkeypox infection** Monkeypox is a zoonotic disease (transmitted from animals). Within 12 days of contracting the virus, infected individuals develop fever, headache, muscle aches, backache, swollen lymph nodes and exhibit rash. The rash develops into vesicles, which eventually become crusty, scab over, and fall off. The disease is transmitted via contact with monkeys or animals that have been in contact with monkeys and can also spread from person-to-person through respiratory droplets or through body fluids. There is no recommended treatment for this infection.

**3.1.6.ic. Smallpox infection (variola virus)** The ordinary cases of smallpox are characterized by the appearance of lesions on the oral mucosa or palate and spread to the face, forearms, hands and finally to the rest of body (Fig. 5). The disease is associated with high fever, malaise, headache, vomiting. As in above, skin spots progress through stages of macules,

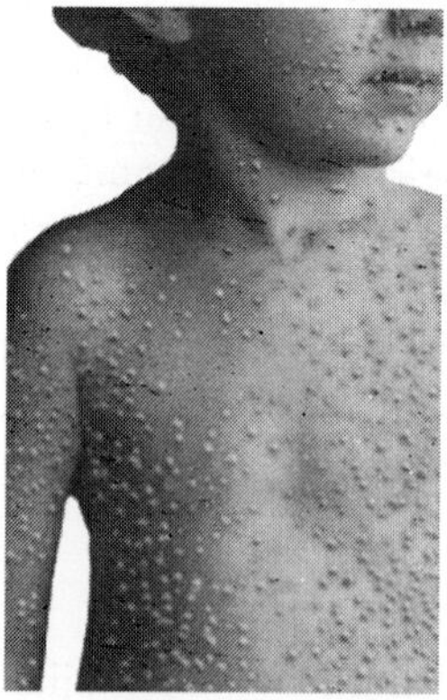

**Fig. 5** Smallpox-infected patient (source: www.nlm.nih.gov/medlineplus). *Colour image of this figure appears in the Colour Plate Section at the end of the Book.*

papules, vesicles and crusts, which fall off within 2 – 4 weeks. About 3% of cases develop haemorrhagic smallpox (mortality rate of more than 94% in unvaccinated individuals). Smallpox is highly contagious and infected patients must be quarantined. The world was declared free of smallpox by the World Health Organization in 1980.

**3.1.6.ii. Orf infection** Orf is a zoonotic disease (transmitted from animals, i.e., sheep and goat). It is characterized by red spots (usually on the back of hands and fingers), mild fever and malaise for a few days. The red spots undergo different changes and become crusts. The disease is self-limiting and patients are healed within a month. The disease is transmitted through contact with infected sheep and goat (person-to-person transmission has not been observed).

**3.1.6.iii. Molluscum contagiosum infection** After contracting the virus, within a few weeks, it appears as small, pink-coloured spots which appear on the face, arms and legs of infected individuals, especially children (Fig. 6). If the infection is sexually transmitted the spots can be seen on the genitals and thighs. Infection is normally transmitted through direct contact with infected persons or indirectly by sharing towels or from swimming pools. There is no recommended treatment as the disease disappears within a few weeks but may require cryotherapy or removal of spots or topical application of cidofovir in severely infected persons, i.e., immunocompromised patients.

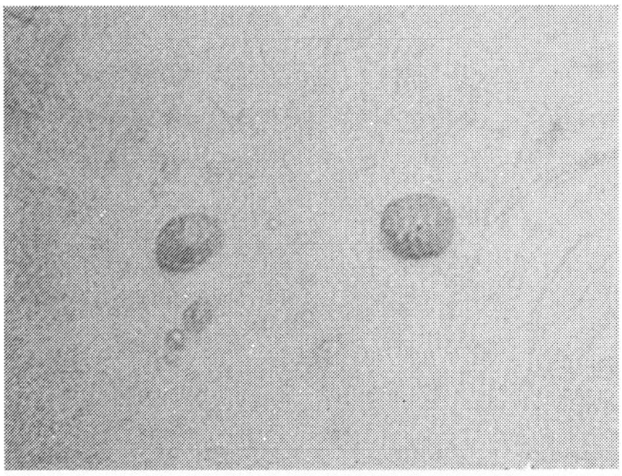

**Fig. 6** Molluscum contagiosum infection (source: www.nlm.nih.gov/medlineplus). *Colour image of this figure appears in the Colour Plate Section at the end of the Book.*

## 3.2 Skin infections – Bacterial

### *3.2.1 Impetigo*

Generally occurs around the mouth, nose or in skin lesions such as insect bites, scratches, skin allergy (Fig. 7). It starts as blisters, which burst to reveal fluid that eventually dries into thick scabs. It may be associated with swollen lymph glands and children under the age of two years may

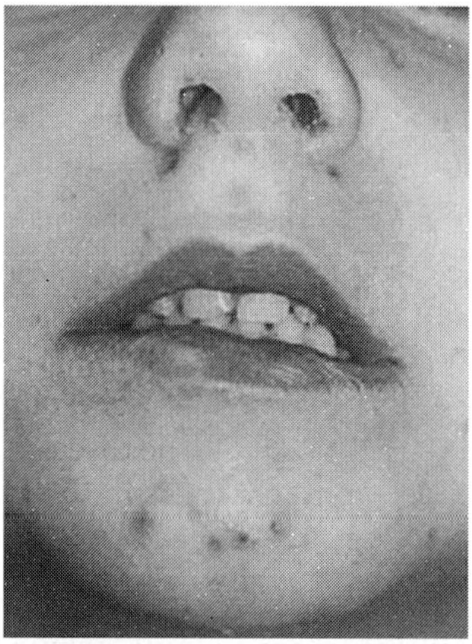

**Fig. 7** Impetigo-infected child (source: www.nlm.nih.gov/medlineplus).
*Colour image of this figure appears in the Colour Plate Section at the end of the Book.*

suffer from fever, diarrhoea and general weakness. The disease spreads by contact with infected individuals or through the use of contaminated objects. It is normally treated with the topical application of antibiotics (for 2 – 3 days) or a complete course of oral antibiotics in severe cases. The usual antibiotics are nafcillin (a penicillinase-resistant synthetic penicillin), flucloxacillin or cloxacillin, while cephalosporins, rifampicin or trimethoprim are also effective. It is caused by Gram-positive bacteria (usually *Staphylococcus aureus* and *Streptococcus pyogenes*).

### 3.2.2 Furunculosis (also called boils or furuncles)

This is an infection of hair follicles and is presented with red spots or pustules centred on a hair follicle. If these lesions become large, they form abscesses (pus-filled cavity resulting from bacterial infection) and may cause fever and malaise. The treatment involves the use of antibiotics such as nafcillin, flucloxacillin or cloxacillin, while cephalosporins, rifampicin or trimethoprim are also effective. It is caused by Gram-positive bacteria (usually *Staphylococcus aureus* and *Streptococcus pyogenes*).

### 3.2.3 Scalded skin syndrome (in neonates it is called Ritter's or Lyell's syndrome)

It is an acute exfoliation (shedding of the outer layer) of the skin. Usually presented as red rash (patchy or localized) but in severe cases, the whole surface of the skin exfoliates exhibiting a scalded appearance (such as a burn caused by hot liquid). The disease is associated with fever, malaise, tenderness and dehydration. The treatment as given above is effective. It is usually caused by *Staphylococcus aureus*.

### 3.2.4 *Erysipelas*

This is a severe skin rash accompanied by fever and vomiting (Fig. 8). Bacteria are initially inoculated into an area of the skin lesion and become red which may develop into numerous vesicles. The disease is associated with malaise, chills, high fever within 48 h of the cutaneous involvement. The infection rapidly spreads through the lymphatic vessels and produce skin swelling and tenderness. The treatment involves topical application of antibiotics as well as the oral antibiotics such as amoxicillin, ampicillin, flucloxacillin, erthyromycin or in severe cases intravenous treatment of benzylpenicillin, cefuroxime. Recently, roxithromycin and pristinamycin were found to be very effective. It is caused by Gram-positive bacteria (usually *Streptococcus pyogenes*).

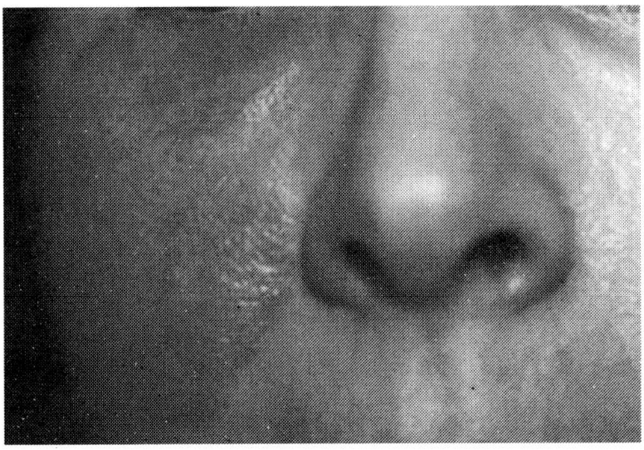

**Fig. 8** Erysipelas-infected patient (source: www.nlm.nih.gov/medlineplus). *Colour image of this figure appears in the Colour Plate Section at the end of the Book.*

### 3.2.5 *Cellulitis*

It is an acute infection of the subcutaneous tissue with inflammation of the overlying skin. The infected skin tissue is presented as red, irritating and painful and more common on the face and lower legs (Fig. 9). Infection initiates with some sort of skin trauma such as an insect bite or lesions and becomes enlarged with the appearance of rash. The disease is associated with fever, chills, warm skin, sweating, tiredness, muscle aches, malaise, nausea, vomiting and hair loss at the site of infection. The treatment involves the use of antibiotics such as cloxacillin, flucloxacillin, cephalosporins or trimethoprim. It is caused by Gram-positive bacteria (usually *Streptococcus pyogenes* and *Staphylococcus aureus*).

### 3.2.6 *Necrotizing infections of the skin and soft tissue*

**3.2.6i Gas gangrene** Gas gangrene is a subcutaneous infection and is a form of decay and tissue death as a result of lack of blood to the area. It generally occurs at the site of trauma or wound with an inflammation

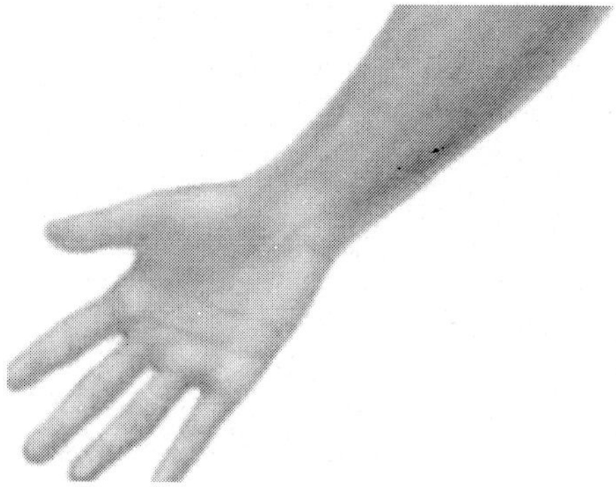

**Fig. 9** Skin lesions due to cellulitis (source: www.nlm.nih.gov/medlineplus).
*Colour image of this figure appears in the Colour Plate Section at the end of the Book.*

followed by painful tissue swelling that form vesicles (also filled with gas), which eventually appear as dark red or purple and result in tissue death. Drainage from the tissue has a foul-smell with brown-red or bloody fluids. The disease is characterized by sweating, fever, anxiety and may develop into increased vascular permeability and hemolysis (destruction of blood). If untreated, patients may suffer from shock (hypotension), renal failure, coma and death. Removal of infected tissue is necessary to control the infection. The treatment includes intravenous application of antibiotics such as benzylpenicillin and metronidazole together with oxygen therapy. Normally caused by Gram-positive bacteria (usually by several species of *Clostridium* spp. but can also be caused by *Staphlococcus aureus*).

**3.2.6ii Necrotizing fasciitis** It is severe type of soft tissue infection that involves the skin and subcutaneous and muscle fat causing gangrene as above (Fig. 10). Infection initiates with bacterial entry into the body through a wound or trauma and appear as painful red spots, which change to purple vesicles and finally necrotic tissue (appear black) with severe pain and swelling in the area. The disease is characterized by fever, sweating, chills, nausea, dizziness, weakness and finally shock. Fourneir's gangrene is necrosis of perineal skin (the region of the skin surrounding the urogenital and anal openings), which may leave the testicles denuded (strip bare). The treatment includes clindamycin, benzylpenicillin together with metronidazole. It is usually caused by Gram-positive bacteria (usually *Streptococcus pyogenes*, also known as 'flesh-eating bacteria'). Skin grafts may be required after the infection is cleared.

### *3.2.7. Otitis externa (also called Swimmer's ear)*

This is an irritating infection of the outer ear that may involve the middle ear (otitis media). It is most common among adults and is acquired by

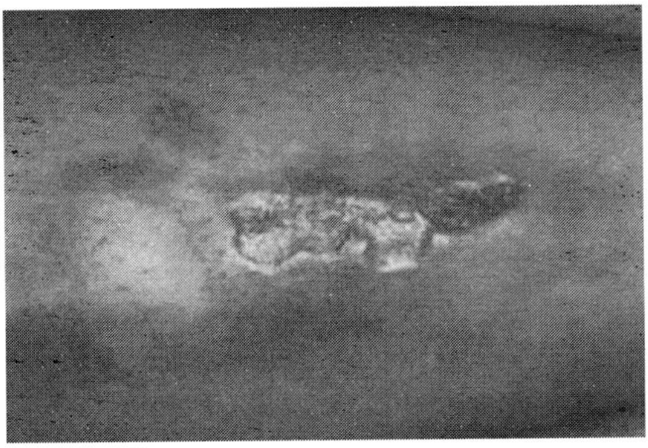

**Fig. 10** Skin lesions due to necrotizing fasciitis (source: www.nlm.nih.gov/medlineplus). *Colour image of this figure appears in the Colour Plate Section at the end of the Book.*

swimming in polluted water or cleaning the ear with a contaminated object. The disease is characterized by ear pain, itching of the ear, pus-like or foul-smell drainage from the ear. The cleaning of the ear may be sufficient to allow resolution but topical application of neomycin, framycetin or clioquinol will help. If infection is due to fungus, clotrimazole is effective. It is usually caused by water-borne pathogens such as *Psedomonas* spp. and occasionally by the fungus, *Aspergillus*.

### 3.2.8 *Erythema chronicum migranes (also called cutaneous lyme disease)*

The infection initiates with a tick bite and exhibits a red rash at the bite place that may cover large areas. It can lasts from a few days to more than a month and is usually associated with pain, itchy with symptoms such as flu-like symptoms, malaise, fever, fatigue, headaches, muscle ache and pains. If not treated, it can involve the cardiovascular and the central nervous system. The treatment involves antibiotics such as doxycycline, erthyromycin or penicillin. It is caused by different species of *Borrelia* spp.

### 3.2.9 *Erythrasma*

It is an inflammation of the skin with red patches and usually occurs in the groin, armpits, and skin folds. It is not painful but is mildly itchy. Topical application of erythromycin is highly effective. It is caused by *Corynebacterium minutissimum*.

### 3.2.10 *Erysipeloid*

This is a skin infection which appears as swollen and red both on hands and on the face (normally seen in individuals who handle meat, as it arises from a zoonotic organism). The symptoms may include fever and chills. The treatment includes penicillin or tetracycline. It is caused by Gram positive *Erysipelothrix rhusiopathiae* (a zoonotic organism that normally causes infection in pigs).

### 3.2.11 *Cat-scratch disease*

This is a bacterial disease that is transmitted to humans by cat scratches or exposure to cat saliva. The infection initiates with papule at the site of injury followed by swelling of the lymph nodes within a few weeks. The disease is characterized by fever, fatigue, malaise, headache, and patients may suffer from weight loss, splenomegaly (enlarged spleen) and a sore throat. It is self-limiting but treatment with tetracycline or erythromycin is recommended in severe cases. It is caused by *Bartonella henselae.*

### 3.2.12 *Actinomycosis*

This is an infection of the skin and subcutaneous tissue usually on the face and neck producing abscesses. Infection normally initiates from the mouth (with injury) or tissue surgery and forms an abscess with reddish lumps on the cheek or upper neck. The disease is characterized by fever, weight loss and is usually painless. The treatment involves the use of penicillin and erythromycin for at least 1 – 2 months. It is normally caused by Gram positive *Actinomyces israelii.*

### 3.2.13 *Cutaneous tuberculosis*

It is characterized by lesions, which normally appear on the hands involving the skin and subcutaneous tissue, which develop into ulcers and discharge pus. It is normally caused by *Mycobacterium marinum.* Other cutaneous infections due to *Mycobacterium* are lupus vulgaris, a granulomatous (tubercle) lesion that develops over years. The reatment involves the application of rifampicin plus ethambutol or clarithromycin. It is caused by different species of *Mycobacterium* spp.

### 3.2.14 *Acne*

It is a common skin infection in young adults and is characterized by small, red spots which turn into pimples (Fig. 11). The inflammation is self-limiting and disappears within a few days. In severe cases, it leaves scars. The use of doxycycline or erthyromycin is effective. It is caused by *Propionibacterium acnes.*

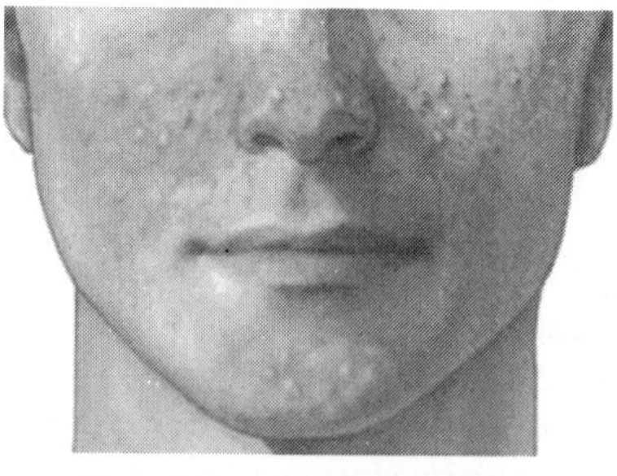

**Fig. 11** Acne, a common skin infection (source: www.nlm.nih.gov/medlineplus). *Colour image of this figure appears in the Colour Plate Section at the end of the Book.*

## 3.3 Skin Infections – Fungal

### *3.3.1 Candidiasis of the skin*

It is the infection of warm, moist areas of the skin such as under the breasts, under the abdomen, the skin beneath an infant's diaper or oral thrush (white patches in the mouth) (Fig. 12). It is characterized by rash, red patches with papules and vesicles. The treatment includes topical application of anti-fungal agents such as nystatin or amphotericin, while itraconazole, fluconazole, imidazole are also effective. It is normally caused by *Candida albicans*.

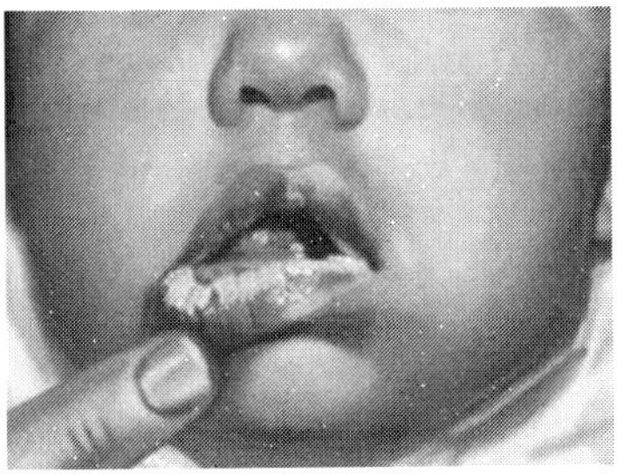

**Fig. 12** Candidiasis of the skin (source: www.nlm.nih.gov/medlineplus).
*Colour image of this figure appears in the Colour Plate Section at the end of the Book.*

### *3.3.2. Dermatophytoses (also known as tinea infections)*

It is generally characterized by an expanding lesion with inflamed advancing edges and is described by its position, tinea capitis (scalp), tinea corporis (body) or tinea cruris (groin) as well as tinea of the nails, tinea of the foot (Athlete's foot) and tinea versicolor (Fig. 13). The use of miconazole, clotrimazole, itraconazole or fluconazole is effective against tinea infections. These are normally caused by *Candida* spp., *Microsporum* spp., *Trichophyton* spp., or *Pityrosporum ovale*.

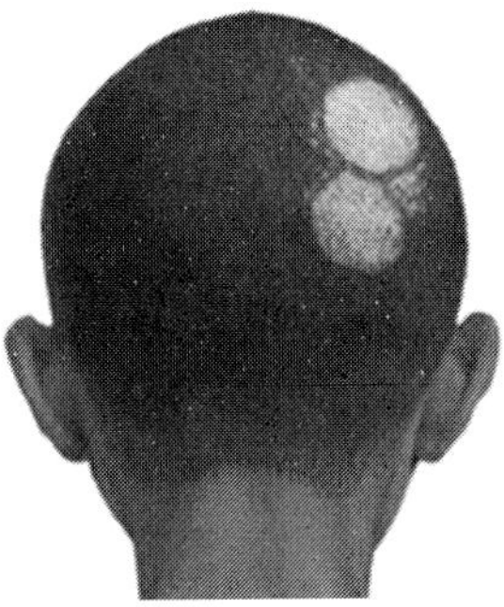

**Fig. 13** Tinea infection (source: www.nlm.nih.gov/medlineplus).
*Colour image of this figure appears in the Colour Plate Section at the end of the Book.*

### *3.3.3 Sporotrichosis*

This is a chronic skin infection (long-term) that initiates when the skin is broken during handling vegetables or plant material. It is characterized by small, red spots that appear at the site of infection and ulcerate (Fig. 14). These lesions do not heal unless treated. The pathogen disseminates in immunocompromised patients and may cause arthritis or even meningitis. The treatment involves the use of potassium iodide for several months. The antifungal agents, itraconazole or ketoconazole are also beneficial. Disseminated infections require Amphotericin B application in addition to others. It is caused by *Sporothrix schenckii.*

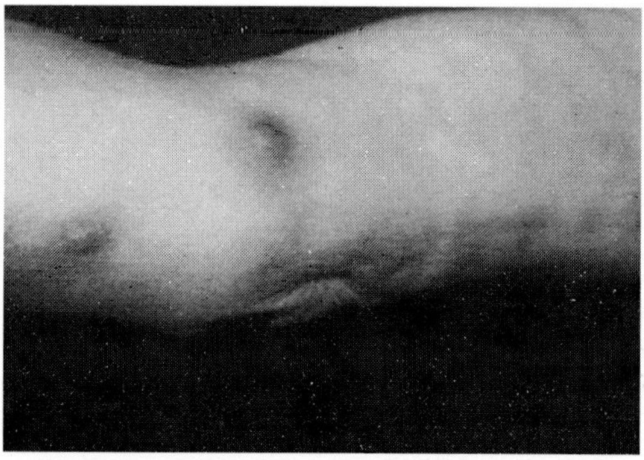

**Fig. 14** Sporotrichosis-infected patient (source: www.nlm.nih.gov/medlineplus). *Colour image of this figure appears in the Colour Plate Section at the end of the Book.*

## 4. EYE INFECTIONS

The eye is covered by eye lids that keep the surface lubricated and free of debris, the mucous membrane (also called conjunctiva) which covers the inner eye and lines the inner side of the eye lids, containing sweat glands to keep the eye surface lubricated. In addition, the eye contains lacrimal gland (tear gland) that secrete enzymes to attack microbial pathogens (Fig. 15).

## 4.1 Eye Infections – Viral

### *4.1.1 Conjunctivitis (also called pink eye)*

Conjunctivitis is the infection of the membrane lining the eye lids (conjunctiva). During common colds, influenza and measles, conjunctiva may become inflamed. The eye feels sore, itchy, appears red, is sensitive to light with watery or mucoid material secretions. The eye may become dry during sleep and the eye lids may be glued together with swelling. It is a self-limiting disease and resolves within a few days. The infection spreads

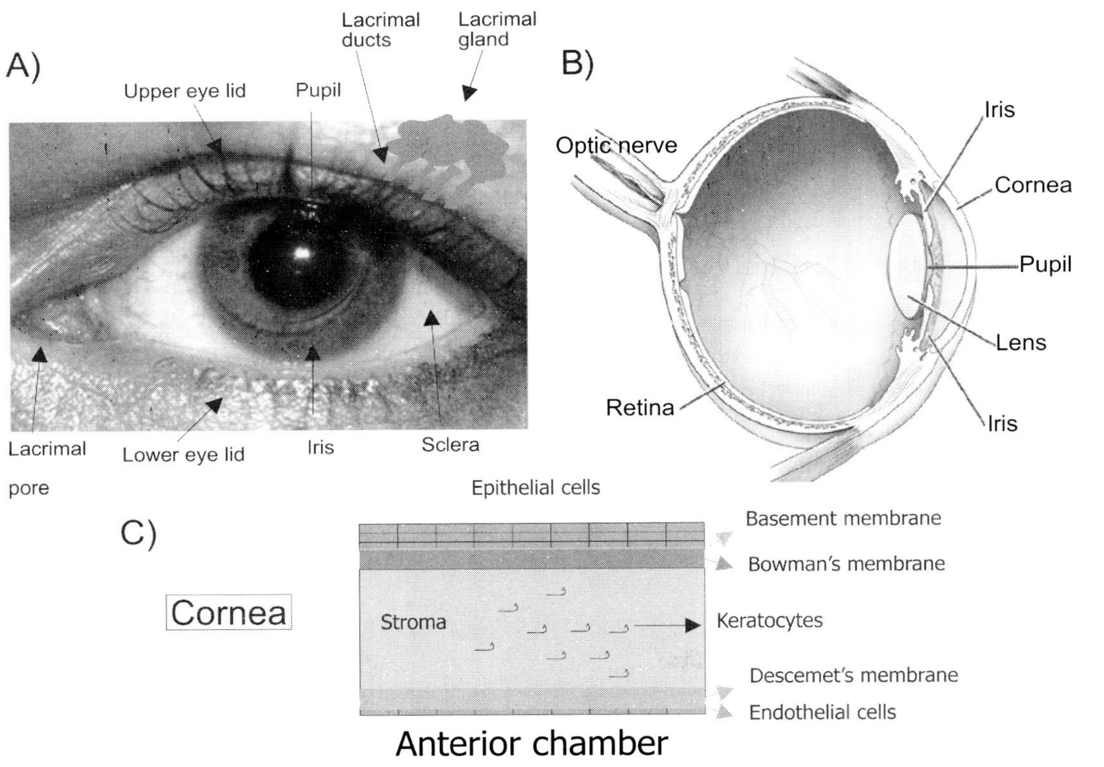


**Fig. 15** (A and B) Anatomy of the eye and (C) cellular structure of cornea.
*Colour image of this figure appears in the Colour Plate Section at the end of the Book.*

between eye to eye or contact with a contaminated instrument. Predominant causative agents are Adenovirus, Enterovirus or Herpes simplex virus.

#### *4.1.2 Keratitis*

It is the infection of the corneal epithelium and is presented with red and painful eyes and exhibited as a branching ulcer. If not treated, the infection may lead to extensive corneal epithelial damage and damage of the underlying tissue, resulting in vision impairment. The disease is characterized by pain, impaired vision, photophobia and watery eyes. The treatment involves the use of antiviral agents, cidofovir and acyclovir. The usual causative agent is Herpes simplex virus.

## 4.2 Eye Infections – bacterial

#### *4.2.1 Conjunctivitis (also called bacterial conjunctivitis)*

As above, it is the infection of the membrane lining the eye lids. This is a serious disease and is characterized by tearing, redness, sore, itchy, blurred vision with watery or purulent (pus-like) discharge. The correct identification requires culturing of the bacteria. The treatment involves the use of chloramphenicol or gentamicin together with neomycin. The causative agents are *Haemophilus influenzae, Streptococcus pneumoniae* and rarely *Staphylococcus aureus, Moraxella lacunata* and *Pseudomonas aeruginosa*.

#### *4.2.2 Neonatal conjunctivitis*

It is a red eye infection of the conjunctivitis of the newborn. The disease is characterized by watery, bloody or pus-like drainage from the infant's eye and swollen eye lids. The treatment includes the use of antibiotic, benzylpenicillin, chloramphenicol or cefotaxime is effective.

The causative agent is *Neisseria gonorrhea* and *Chlamydia trachomatis,* which are usually passed from mother to child during birth, however occasionally viruses such as herpes can also cause this infection.

#### *4.2.3 Trachoma*

The repeated infection of *Chlamydia trachomatis* due to crowding or poor hygiene (unwashed hands) results in the swelling of the eye lids (or even deformity), scarring, cloudy cornea and if not treated may lead to blindness. The disease spreads from the infected eye secretions or nasal secretions. The treatment includes the use of chlortetracycline or doxycycline which is effective.

#### *4.2.4 Keratitis*

The symptoms are similar to the above with purulent discharge. If not treated, the infection may lead to extensive corneal damage with blinding consequences. The treatment involves the use of antibiotics, gentamicin together with neomycin. If the infection involves both the cornea and conjunctiva, it is called keratoconjunctivitis. The usual causative agent is *Pseudomonas aeruginosa*.

## 4.3 Eye Infections – Protozoa

### *4.3.1 Keratitis*

The symptoms are similar to the above except that the infection is associated with extremely intense pain. A ring infiltrate is often observed in the stroma. Early and aggressive treatment is crucial for a successful outcome. The treatment involves topical application of pentamidine isethionate, chlorhexidine, 'polyhexamethylene biguanide' together with neomycin. The usual causative agent is *Acanthamoeba* spp.

# 5. EAR INFECTIONS

The ear is composed of three regions, (i) external ear which is visible to the naked eye and collects sound waves and directs them to the eardrum, (ii) middle ear which collects and amplifies sound waves and transmits them to the inner ear, and is connected to the pharynx (throat), and (iii) inner ear that contains sensory organs and receptors responsible for hearing.

## 5.1 Ear infections – Bacterial

### *5.1.1 Acute otitis media*

It is an inflammation of the middle ear, a very common and painful infection of the ear predominantly in infants and children because their eustachian tubes (runs between the middle ear and back of the throat to keep the flow of liquid) become obstructed easily and result in fluid build up, which becomes infected (especially in children who drink while lying on their back). The disease is characterized by intense pain and may exhibit purulent discharge and is associated with fever. The treatment is normally focussed on pain (pain relief drops and ibuprofen or acetaminophen) together with antibiotics, amoxicillin and neomycin.

The causative agents are normally *Haemophilus influenzae*, *Streptococcus pneumoniae*, *S. pyogenes* and *Staphylococcus aureus*.

# 6. RESPIRATORY INFECTIONS

The respiratory tract can be divided into two parts, (i) upper respiratory tract that includes the nose, paranasal sinuses, pharynx (throat) and consists of nasopharynx, oropharynx and laryngopharynx, larynx (voice box), and (ii) lower respiratory tract which includes the trachea (windpipe), lungs containing bronchi, bronchioles and alveoli which import oxygen into the body and export carbon dioxide out of the body (Fig. 16).

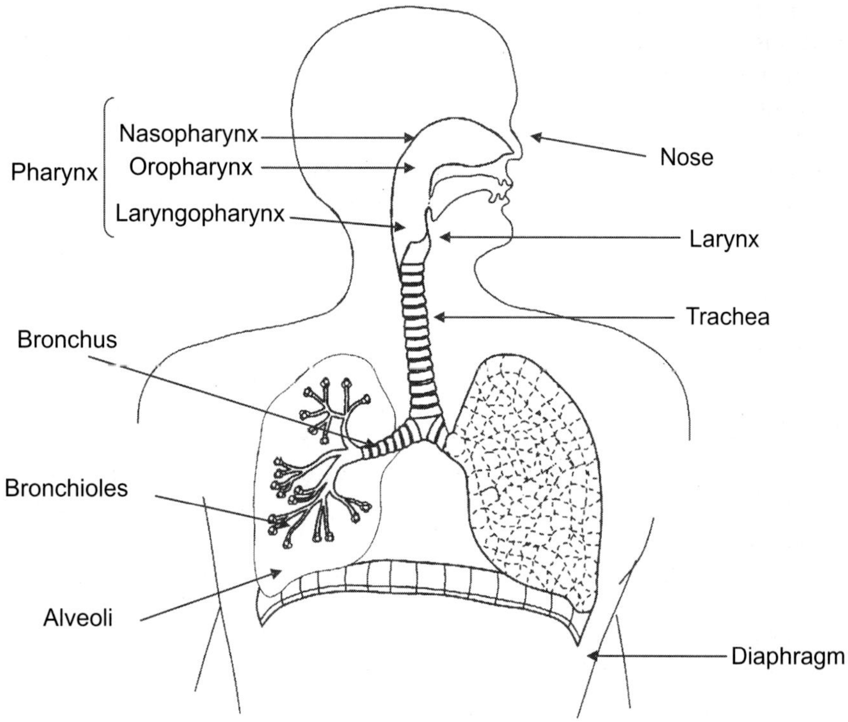


**Fig. 16** Respiratory system.

## 6.1 Upper Respiratory Infections – Viral

### *6.1.1 Paranasal sinusitis*

In this infection, sinuses become blocked and filled with exudates, which become infected. The disease is characterized by pain and tenderness over the affected sinuses and an x-ray shows thickening of the soft-tissue of the cavity. The treatment should include decongestants together with antibiotics such as co-amoxiclav or tetracycline. The causative agents are normally *Haemophilus influenzae*, *Streptococcus pneumoniae* and *Staphylococcus aureus*.

### *6.1.2 Common cold*

It is characterized by a runny nose, nasal congestion, sneezing, sore throat, cough and headache. Secretions are full of 'cold' (viruses) and spread easily between individuals and also to other tissues (touching eye, mouth, etc.). Adults normally do not develop fever but young children develop high fever and this may lead to ear infections. It is usually self-limiting within few days but excessive hot fluids with salts should be taken. More than 200 viruses can cause the common cold.

### 6.1.3 *Severe acute respiratory syndrome (SARS)*

SARS is a viral pneumonia, with symptoms including fever, dry cough, breathing difficulty, headache and hypoxaemia (low blood oxygen concentration). Death may occur due to progressive respiratory failure and because of alveolar damage. The patient may feel better following the first few days of infection, but worsening during the second week, which may be due to the patient's immune responses rather than uncontrolled viral replication. There is no specific treatment and supportive care is helpful. The causative agent is coronovirus which is transmitted via respiratory droplets during sneezing or coughing.

### 6.1.4 *Pharyngitis*

It is an inflammation of the pharynx. The disease is characterized by a sore throat, discomfort in swallowing, fever, swollen lymph nodes in the neck and muscle aches. Negative culture for bacteria should be suspected for viral pharyngitis. There is no recommended treatment but gargling with warm, salted water should be helpful and anti-inflammatory drugs together with antiviral, pleconaril in severe cases (usually resolves within days). The normal causative agent is influenza virus as well as adenovirus, which causes a severe sore throat and usually occurs in winter. During the summer time, viral pharyngitis may be due to enteroviruses (including polioviruses, coxsackieviruses, echoviruses), which may involve the skin and the CNS.

### 6.1.5 *Mononucleosis (also known as kissing disease)*

The disease is characterized by fever, a sore throat, swollen lymph glands in the neck, malaise, muscle aches, rash, enlarged tonsils and an enlarged liver or spleen. As suggested by the name, it is spread by saliva. There is no recommended treatment but ibuprofen can be used for fever and pain (resolves within a month). The causative agents are normally Epstein-Barr virus (EBV) and rarely cytomegalovirus (CMV), members of Herpesvirus family.

### 6.1.6 *Measles (childhood infections)*

It is characterized by fever, cough, runny nose, fever, a sore throat, inflammation of the eyes, photophobia, and rash appearing within a few days of contracting the virus. There is no specific treatment but the use of MMR vaccine (measles, mumps, and rubella) is effective in preventing this infection. The causative agent is the measles virus which is highly contagious and spreads via respiratory droplets from infected persons.

### 6.1.7 *Mumps (childhood infections)*

It is a painful enlargement or swelling of the salivary glands associated with pain on the face, fever, headache, a sore throat and swelling of the jaw. There is no specific treatment but gargling with warm, salted water and corticosteroid relieves the pain. The use of MMR vaccine is effective in preventing the infection. The causative agent is the mumps virus which is highly contagious and spreads via respiratory droplets from infected persons.

### *6.1.8 Rubella (childhood infections)*

The disease is characterized by fever, headache, malaise, aches, runny nose, rash and very rarely encephalitis. There is no specific treatment but the use of acetaminophen (for fever) and anti-inflammatory agents can help relieve symptoms. The use of measles–mumps–rubella (MMR) vaccine is effective in preventing the infection. The causative agent is rubella virus which is transmitted through the air or close contact with the infected person or from an infected mother to the foetus, which can result in serious foetal defect.

## 6.2 Upper Respiratory Infections – Bacterial

### *6.2.1 Tonsillitis*

It is an inflammation of tonsils and is characterized by a sore throat, difficulty in swallowing, headache, fever, chills, loss of voice, throat tenderness and visible enlarged tonsils with intense pain. It may also be present in individuals suffering from pharyngitis. The treatment involves the use of antibiotics such as penicillin or benzylampicillin and excessive fluidintake and gargling with warm, salted water. The causative agents are normally *Streptococcus pyogenes, Haemophilus influenzae, Corynebacterium diphtheriae, Streptococcus pneumoniae* and *Staphylococcus aureus*.

### *6.2.2 Scarlet fever*

It is characterized by a sore throat, fever, vomiting, rash, peeling of finger tips, toes and the groin as the rash disappears, chills, abdominal pain and malaise initially with the tongue appearing furred. The treatment includes the use of penicillin or benzylampicillin which should clear the infection within 2 – 3 weeks. The usual causative agent is *Streptococcus pyogenes*.

### *6.2.3 Epiglottitis*

It is an inflammation of the cartilage which covers the trachea (windpipe) and is most common in children. The disease is characterized by fever, sore throat, chills, breathing difficulty as well as difficulty in swallowing with enlargement of the epiglottitis. The treatment involves the use of cephalosporins, oxygen therapy and anti-inflammatory drugs. It is usually caused by *Haemophilus influenzae*.

### *6.2.4 Diphtheria*

It is characterized by a sore throat with painful swallowing, fever, chills, bloody nasal secretions, cough, enlarged lymph glands in the neck, breathing difficulty, skin lesions may be present and may spread via the bloodstream to other organs including the CNS. The disease is transmitted with infectious respiratory droplets or contaminated objects. The treatment includes the use of antibiotics, benzylpenicillin, erythromycin or cephalosporin and oxygen therapy. The disease is caused by diphtheria toxin so treatment involves antitoxin to neutralize the toxin. It is to be noted that the protective effect of tetanus vaccine lasts only 10 years. The causative agent is *Corynebacterium diphtheriae.*

#### 6.2.5 *Ludwig's angina*

It is an inflammation of the floor of the mouth (under the tongue) which blocks the airway. It is characterized by swelling of the neck with pain, breathing difficulty, fever, malaise, confusion and changes in mental status. It is usually secondary to tooth root infection or following mouth injury. The treatment includes the use of penicillin together with flucloxacillin or cephalosporin together with metronidazole in severe cases. The causative agents are usually *Streptococcus pyogenes, Staphylococcus aureus* and rarely *Prevotella melaninogenicus.*

#### 6.2.6 *Retropharyngeal abscess*

It is characterized by a sore throat, tooth abscess, fever, swallowing and breathing difficulty with characteristic pus at the back of the throat usually in children. The treatment involves drainage of the abscess and includes the use of clindamycin together with metronidazole in severe cases. The causative agents are usually *Streptococcus pyogenes, Staphylococcus aureus* and rarely *Prevotella melaninogenicus.*

## 7. LOWER RESPIRATORY INFECTIONS

These include infections of the trachea (windpipe), bronchi and bronchioles.

### 7.1 Lower Respiratory Infections – Viral

#### *7.1.1 Croup*

Croup is a swelling around the vocal cords resulting in breathing difficulty and is usually restricted to children. It is typically characterized by a 'barking cough' which lasts for a few days with a high-pitched whistling sound produced by air flowing through the narrowed breathing tubes. However, in severe cases it can last for weeks with serious consequences. It is normally self-limiting (recovery within a week) and treatment is to relieve symptoms including a warm, steamy atmosphere together with dexamethasone (a corticosteroid that relieves the inflammation). Croup is normally caused by parainfluenza viruses, influenza virus, measles virus, adenovirus and respiratory syncytial virus, however the involvement of bacteria (e.g., *Haemophilus influenzae, Corynebacterium diphtheriae*) requires antibiotic treatment.

### 7.1.2 *Bronchiolitis*

It is an inflammation of the bronchioles (small air way branches in the lungs), usually limited to infants and children. It is characterized by short, rapid breathing with a cough. The shortness of breath makes it hard to maintain breathing. As for croup, it is self-limiting (recovery within a week) and treatment is targetted against symptoms with humidified air and even oxygen therapy in severe cases. The use of dexamethasone or antiviral agent, ribavirin is effective. Respiratory syncytial virus (RSV) is a common cause, which is transmitted from person to person through contact with nasal secretions. Other viruses, parainfluenza viruses, influenza virus and adenovirus may also cause this disease.

### 7.1.3 *Influenza (also called flu)*

It is an infection of the respiratory tract and associated with damage to the epithelial lining and is highly contagious. The disease is characterized by a sore throat, fever, dry cough, runny nose, chills, malaise, aches, tiredness which appear within a day of contracting the virus and can lasts for up to a week. The disease can be serious for people over the age of 50, very young children or immunocompromised patients. It is different from the cold and can only infect a person once every few years, while a cold can infect the same person several times a year. The disease is highly contagious and is transmitted via droplets from infected individuals while sneezing or coughing. It is self-limiting (recovery within a week) and treatment is focused on the symptoms and includes rest, intake of excessive liquids, however antiviral agents such as amantadine, rimantadine, oseltamivir and zanamivir are also effective. The common causative agents are Influenza A, Influenza B and Influenza C.

## 7.2 Lower Respiratory Infections – Bacterial

### 7.2.1 *Chronic obstructive pulmonary disease (also called chronic bronchitis)*

This is an inflammation of the bronchi (the tube that carries air from the windpipe into the lungs) together with a cough and excessive mucous production, impairing air flow to the lungs. It is a long-term disease lasting for several months and even years and is usually limited to smokers and the elderly. Patients suffer from shortness of breath, a cough with mucous, fatigue, chest discomfort, limited ability to walk or do any exercize with swelling of the feet and legs. The treatment involves the use of corticosteroids together with antibiotics, tetracycline, ampicillin, amoxicillin, clarithromycin or co-amoxiclav. The common causative agents are *Streptococcus pneumoniae*, *Haemophilus influenzae* and *Moraxella catarrhalis*.

### 7.2.2 *Cystic fibrosis*

It is an inherited disease associated with the buildup of thick, sticky mucus in the lungs and the digestive tract. The defective gene induces the body to produce thick mucus which blocks the bronchi and the resident bacterial flora can cause infection with serious consequences. Symptoms are highly variable and may include coughing, breathing difficulty, weight loss, diarrhoea, fatigue and recurrent respiratory infections. The treatment is long-term and includes the use of antibiotics (for lung infections), azithromycin and amoxicillin together with pain killers (ibuprofen) and physiotherapy and postural drainage (performing different positions to clear the mucus in the lungs).

### 7.2.3 *Pneumococcal pneumonia*

It is an inflammation of the lungs and is associated with a cough, with yellowish mucus, fever, chills, chest pain, breathing difficulty, fatigue and excessive sweating. The treatment includes the use of antibiotics, benzylpenicillin, amoxicillin or co-amoxiclav. The common causative agent is *Streptococcus pneumoniae* (pneumococcus).

### 7.2.4 *Legionnaires' disease*

This is a respiratory infection that can vary from a mild cough to serious pneumonia. The disease is characterized by a cough (possibly with blood), breathing difficulty, chest pain, headache, fever, malaise and body aches with fatal consequences, especially in the elderly. The treatment includes immediate use of antibiotics, ciprofloxacin, azithromycin, clarithromycin or erythromycin and may require oxygen therapy. The causative agent is *Legionella pneumophila*.

### 7.2.5 *Atypical pneumonia*

It is known as atypical as this form of mild pneumonia differs from acute pneumonia. It is characterized by mild symptoms such as prolonged fever, malaise and sweating. The treatment includes the use of antibiotics, erythromycin, tetracycline, clarithromycin or chloramphenicol. The common causative agents are *Mycoplasma pneumoniae, Coxiella burnetii* and *Chlamydophila pneumoniae*.

### 7.2.6 *Q fever (also known as Query fever)*

This is a form of an atypical pneumonia caused by *Coxiella burnetii*. This bacterium can also cause liver inflammation (hepatitis). The bacterium normally affects farms animals and is usually transmitted to humans by contact with animals or consumption of raw milk. The disease is characterized by a flu-like illness, followed by pneumonia and hepatitis. The treatment includes the use of antibiotics, erythromycin, tetracycline, clarithromycin or chloramphenicol. The common causative agent is *Coxiella burnetii*.

### *7.2.7 Pertussis (also known as whooping cough)*

It is a serious disease associated with violent coughing. The disease is characterized by a runny nose, fever, vomiting, breathing difficulty and violent coughing appearing within 10 days of contracting the disease and can have serious consequences. The treatment includes the use of erythromycin and amoxicillin with oxygen therapy in severe cases. The common causative agent is *Bordetella pertussis.*

## 7.3 Lower Respiratory Infections – Fungal

### *7.3.1 Aspergillosis*

It is an allergic reaction to *Aspergillus* fungus, which is commonly found in decaying vegetation, stored grains, etc. The disease is characterized by fever, cough, (possibly with blood), breathing difficulty with severe symptoms in asthma sufferers (called pulmonary aspergillosis) (Fig. 17). It may invade other organs causing bone pain, blood in urine, endocarditis and meningitis and this form is called 'invasive aspergillosis'. If the disease is limited to the lungs, then treatment is not necessary but for severe cases it includes the use of itraconazole or amphotericin B. The causative agent is *Aspergillus* spp.

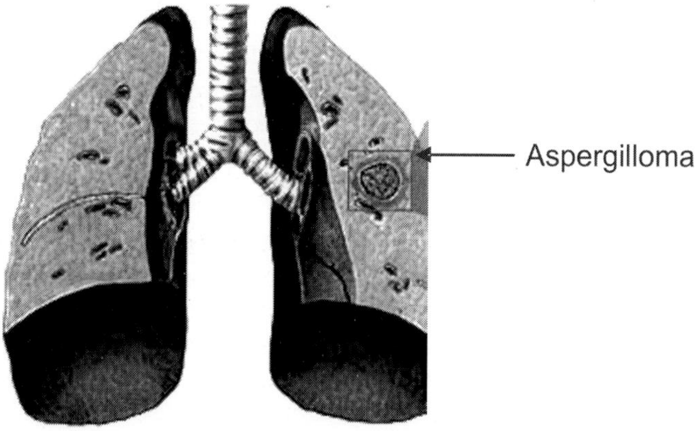


**Fig. 17** *Aspergillus*-infected lungs.

### *7.3.2 Histoplasmosis*

It is a fungal infection of the lungs and is usually asymptomatic. Acute clinical disease exhibits a cough, chest pain, fever, chills and lymphadenopathy. This fungus grows in soil and is transmitted to humans via airborne fungal spores. As above, it may invade other tissues with fatal consequences. If the disease is limited to the lungs then treatment is not necessary but in severe cases this includes the use of itraconazole, ketoconazole or amphotericin. The causative agent is *Histoplasma capsulatum* which is usually present in the soil and transmitted to humans via inhalation.

#### *7.3.3 Blastomycosis*

It is a rare infection associated with breathing difficulty, fever, cough, fatigue, malaise, muscular pain, joint pain and malaise. Treatment is not necessary for lung infection but if the disease spreads to other tissues or organs, the use of itraconazole, ketoconazole or amphotericin is recommended. The causative agent is soil/wood borne *Blastomyces dermatitidis*.

#### *7.3.4 Coccidioidomycosis*

Fungal infection of the lungs that may be characterized by a cough, chest pain, fever, chills, headache, body aches, excessive sweating, confusion, stiffness in the neck and malaise. As above, it may invade other tissues with fatal consequences. If the disease is limited to the lungs then treatment is not necessary but in severe cases, this includes the use of ketoconazole, amphotericin B or fluconazole. The causative agent is *Coccidioides immitis* usually found in dust, desert alkaline soils and transmitted to humans via inhalation.

## 8. GASTROINTESTINAL INFECTIONS

The digestive system consists of (i) the digestive tract that begins with the oral cavity and continues through the pharynx (throat), oesophagus, stomach, small intestine and large intestine ending at the anus. The term gastrointestinal tract (GI tract) is normally referred to as the lower digestive tract, i.e., the stomach and intestines. Other accessory organs associated with the digestive tract are the salivary glands (which produce saliva to digest carbohydrates), pancreas (which secrete insulin that reduces glucose in the blood and glucagons that increases sugar levels in the blood), liver (produces bile which consists of water, ions, cholesterol and various other lipids made of cholesterol that enters the small intestine and is required for normal digestion) and the gallbladder (stores excess bile that may be concentrated) (Fig. 18). Overall, the gastrointestinal tract together with accessory organs digest food, absorbs small organic molecules and excrete waste products.

### 8.1 Gastrointestinal Infections – Viral

#### *8.1.1 Primary herpetic gingivostomatitis*

Normally limited to infants but may occur in young children and adults. Only 5 – 10% of the infected patients develop this disease, which is

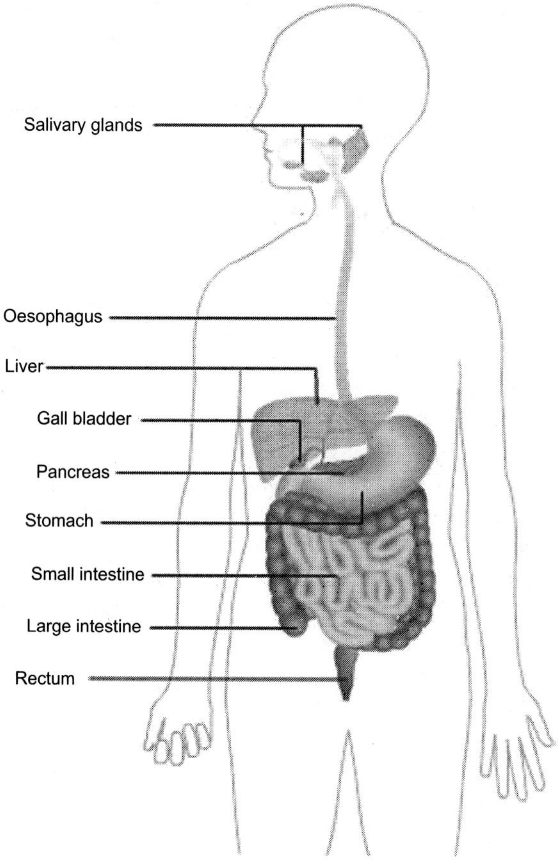


**Fig. 18** Gastrointestinal system.
*Colour image of this figure appears in the Colour Plate Section at the end of the Book.*

characterized by cervical lymphadenopathy, malaise, fever and exhibit whitish vesicles (discrete areas of ulceration) in the mouth, i.e., tongue, palate (roof of the mouth), gums or inside the lips. The symptoms usually last one to two weeks and usual diagnosis is made on clinical presentation. The treatment includes anti-viral drugs, acyclovir (Zovirax®) or valacyclovir. The causative agent is Herpes simplex virus that is transmitted via direct contact with the infected person.

### *8.1.2 Viral gastroenteritis*

It is an inflammation of the stomach and intestines and is characterized by abdominal pain, diarrhoea, nausea, vomiting and may involve fever and chills usually affecting the young and immunocompromised individuals.

The symptoms appear within a few hours of contracting the virus. There is no recommended treatment against the virus but targetted against the symptoms by excessive intake of fluids (with salts and minerals) and the disease resolves within a few days. Anti-diarrhoeal drugs should not be given. The common causative agents are rotavirus (common in children) and Norwalk virus (associated with groups), which are transmitted via contaminated water/food.

## 8.2 Gastrointestinal Infections – bacterial

### *8.2.1 Bacterial gastroenteritis*

It is an inflammation of the stomach and intestines caused by bacteria and bacterial toxins. The disease is characterized by abdominal pain and cramps, diarrhoea, nausea, vomiting and may involve bloody stools. To differentiate from viral disease, the culturing of bacteria from stools is crucial. If the inflammation is an acute inflammation of the stomach, it is called 'acute gastritis'. The treatment involves excessive intake of fluids and is usually self-limiting. To include antibiotics in the treatment regimen, bacteria must be cultured and its antibiotic susceptibility should be determined before the application of appropriate antibiotics is effective. The common causative agents include *Salmonella* (disease called *Salmonella* enteritis, the antibiotic ciprofloxacin and ofloxacin are effective), *Shigella* (called *Shigella* enteritis, the antibiotic ciprofloxacin and ofloxacin are effective), *Staphylococcus* (called Staph aureus food poisoning, self-limiting), *Campylobacter jejuni* (called *Campylobacter* enteritis), *E. coli* (called *E. coli* enteritis, the antibiotic ciprofloxacin or ceftriaxone are effective), *Helicobacter pylori* (causes gastritis and duodenal ulcer, the antibiotic amoxicillin and clarithromycin are effective) which are transmitted via contaminated food and water.

### *8.2.2 Food poisoning*

It is a result of eating food (stored at room temperature for too long, unrefrigerated etc.) contaminated with bacterial pathogens that occurs in one or groups of people. The disease is characterized by nausea, vomiting, abdominal cramps, diarrhoea, fever and general weakness. As above, the treatment involves fluid intake and is usually self-limiting. To include antibiotics in the treatment regimen, bacteria must be cultured and its antibiotic susceptibility should be determined before the application of an appropriate antibiotics is effective. The common causative agents include *E. coli, Salmonella, Shigella, Staphylococcus, Campylobacter jejuni* (the antibiotic erythromycon or ciprofloxacin are effective), *Yersinia* (the antibiotic doxycycline or oxytetracycline are effective), *Vibrio cholerae* (self-limiting but needs excessive fluids), and a very serious agent, *Clostridium botulinum* [that can cause fatal infection and should be treated with Botulism immune globulin (BIG) is effective].

## 8.3 Gastrointestinal Infections – Protozoa

### *8.3.1 Giardiasis*

It is an infection of the small intestine and is characterized by diarrhoea, abdominal pain, swollen abdomen, nausea, vomiting, headache and weakness lasting for a few weeks. The treatment involves the use of metronidazole or quinacrine. The causative agent is *Giardia* spp. that is transmitted to humans via contaminated water.

### *8.3.2 Cryptosporidiosis*

As with Giardiasis, it is an infection of the small intestine and is characterized by watery diarrhoea, nausea, abdominal cramps and malaise. There is no recommended treatment but nitazoxanide or paromomycin are shown to be effective. The causative agent is *Cryptosporidium* spp. which is transmitted to humans via contaminated water.

### *8.3.3 Amoebiasis*

It is an infection of the intestine which is usually asymptomatic but can produce disease characterized by abdominal cramps, bloody stools (7 to 20 per day), fever, vomiting and fatigue. The causative agent is *Entamoeba histolytica* which is transmitted to humans via contaminated water, food or contact through infected individuals. It normally lives in the intestine but can penetrate into the intestinal tissue causing colitis or amoebic dysentery (inflammation of the large intestine) and spreads to other tissues. The treatment includes the use of metronidazole or diiodohydroxyquin.

# 9. LIVER INFECTIONS

## 9.1 Liver Infections – Viral

### *9.1.1 Hepatitis A*

It is an inflammation of the liver (enlarged liver) with similar symptoms to flu but the skin and eyes may also become yellow, i.e., jaundice as bilirubin (a breakdown product of the hemoglobin) is not filtered from the blood as well as dark urine. The symptoms appear within several weeks of contracting the virus. There is no recommended treatment (hepatitis A vaccination is the best option), however the use of corticosteroids may help relieve symptoms. The causative agent is hepatitis A virus (HAV) which is transmitted to humans via contaminated water, food (poor hygiene) or via contact with infected patients.

### 9.1.2 *Hepatitis B*

It is an inflammation of the liver (enlarged liver) with symptoms similar to hepatitis A, but some patients infected with hepatitis B can develop chronic infection which may lead to permanent liver damage or liver cancer. The treatment is similar to hepatitis A for majority of the cases, but for chronic hepatitis B, use of interferon and antivirals, lamivudine and adefovir dipivoxil are effective while the hepatitis B vaccination is the best option. The severity of hepatitis B is increased with the association of hepatitis D virus (also called delta virus). In the majority of cases, hepatitis D only causes infection in association with HBV but the treatment remains the same. The causative agent is hepatitis B virus (HBV) which is transmitted to humans via the blood and other bodily fluids (unprotected sex, blood transfusions, during birth, etc.).

### 9.1.3 *Hepatitis C*

As for others, it is an inflammation of the liver which may be associated with jaundice, abdominal pain, fatigue, weakness, nausea, vomiting and dark urine. There is no recommended treatment for hepatitis C but use of interferon together with antiviral, ribavirin is of value. It rarely causes total liver damage. The causative is hepatitis C virus (HCV), which is transmitted to humans as for HBV.

### 9.1.4 *Hepatitis E*

It can produce an infection similar to hepatitis A and is also transmitted via the foecal-oral route but has a longer time course. Infection can result in liver failure usually in pregnant women. There is no recommended treatment but control requires improved sanitary measures.

### 9.1.5 *Yellow fever*

It is characterized by fever, jaundice (hence the name yellow) and may lead to kidney failure particularly in the elderly. The symptoms develop within a few days of contracting the virus. The disease is characterized by fever, headache, vomiting and jaundice. After a period of recovery, around 15% of infected individual develop serious symptoms including kidney failure, liver damage, haemorrhage, brain dysfunction, seizures, coma, and may lead to death. There is no recommended treatment; however a vaccine is available for people living in endemic areas, South America and Africa. The causative agent is arbovirus which is transmitted by the bite of mosquitoes.

## 9.2 Liver Infections – Bacterial Infections

### 9.2.1 *Leptospirosis*

The disease is characterized by fever, headaches, a dry cough, nausea, vomiting, diarrhoea, and may be associated with joint pain, an enlarged liver and enlarged lymph glands. The treatment involves the use of antibiotics, benzylpenicillins, oxytetracyclines or chloramphenicol. The causative agent is Leptospira (a spirochete bacterium) which is transmitted to humans via water contaminated with animal urine, sewer workers, farmers, etc.

#### *9.2.2 Brucellosis*

It is characterized by flu-like symptoms together with abdominal pain, an enlarged liver, back and joint pains. The treatment includes the use of antibiotics, doxycycline and rifampicin.

The causative agent is *Brucella* which is transmitted to humans via contaminated meat, raw milk, cheese or contact with infected animals (as this bacterium infects pigs, dogs, cattle, goats, camels, etc.).

#### *9.2.3 Pyogenic liver abscess*

It is a liver abscess with a pus-filled cavity. The disease is characterized by fever, chills, nausea, vomiting, weakness, jaundice and dark urine. The treatment includes drainage of the abscess together with the use of antibiotics, penicillin and metronidazole. The common causative agents are *E. coli*, *Klebsiella*, *Enterococcus*, *Staphylococcus* and *Streptococcus*. It can also be caused by appendicitis.

#### *9.2.4 Cholangitis*

This is an infection of the bile duct that carries bile from the liver to the gall bladder and then to the intestine, where it helps in digestion. The disease is characterized by abdominal pain, chills, fever, dark urine, nausea and vomiting. The treatment involves the use of antibiotics, gentamicin, cephalosporin or amoxicillin together with metronidazole. The causative agents are bacteria (*E. coli*, *Klebsiella pneumoniae*, *Clostridium perfringens* and_*Bacteroids fragilis*).

### 9.3 Liver Infections – Protozoa

#### *9.3.1 Amoebic liver abscess*

As above, it is liver abscess with pus and is characterized by fever, abdominal pain, sweating, chills, weakness, diarrhoea, jaundice and joint pains. The treatment includes the use of metronidazole and diiodohydroxyquin. The causative agent is *Entamoeba histolytica* which also causes amebiasis, is acquired through contaminated water, food or through contact with infected individuals and invades the intestinal tissue to gain entry into the blood and carries the blood to the liver.

## 10. URINARY TRACT INFECTIONS

The urinary tract includes the kidney, uterus (the tube that transports urine from the kidneys to the bladder), bladder, and the urethra (the tube that transports urine from the bladder to the outside), with the primary function of excretion of waste products from the blood by urine (Fig. 19).

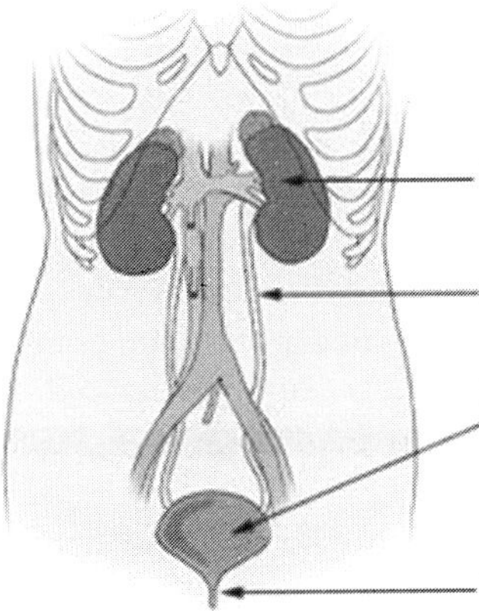


**Fig. 19** Urinary tract system.
*Colour image of this figure appears in the Colour Plate Section at the end of the Book.*

## 10.1 Urinary Tract Infections – Bacterial

### *10.1.1 Cystitis*

It is an inflammation of the lower urinary tract (bladder and urethra) and is characterized by pain or a burning sensation while urination, frequent need of urination, cloudy and smelly urine (possibly bloody), painful sexual intercourse and may also be associated with nausea, vomiting, fever and chills. The treatment includes the use of cephalosporins, amoxicillin or doxycycline and phenazopyridine hydrochloride especially if the kidneys become involved.

The usual causative agents are bacteria (usually *E. coli, Klebsiella pneumoniae*) that enter into the urethra from the anus as they are fairly close especially in women), thus keeping the genitals clean and not dragging bacteria from the rectum to urethra will help control infection. Also, urinating immediately after sexual intercourse and washing will help reduce the risk of infection.

### *10.1.2 Nephropathy (kidney infections)*

It is damage of the kidney and is characterized by abdominal pain, frequent urination, painful or burining urination, dark and possible bloody diarrhoea and may also be associated with vomiting, nausea, fever and chills. Kidney damage due to baceria is known as pyelonephritis.

The treatment requires identification of the causative agent followed by appropriate antibiotics or surgery is needed to relieve symptoms. The causative agents are usually bacteria. This infection may be developed due to the backward flow of urine into the kidney (a condition called reflex neuropathy).

### *10.1.3 Orchitis*

It is the inflammation of one or both the testicles and is charactrized by scrotal swelling, fever, painful urination, pain during intercourse, testicle pain and possibly blood in the semen, resulting in testicular atrophy (shrinking of testicles). The causative agents are bacteria and occasionally viruses (mumps). For bacterial infections, the use of appropriate antibiotics following the correct identification of the organism is effective. For viral causes, treatment may include prednisolone to alleviate inflammation.

## 11. CENTRAL NERVOUS SYSTEM INFECTIONS

The nervous system – the primary function of the nervous system is to monitor internal and external conditions using sensory receptors, co-ordinate the sensory information and direct a response using the other system. It is further divided into:

i. Central nervous system (brain and spinal cord), the brain maintains the memory, intelligence, and emotion. The CNS is responsible for processing sensory information and co-ordinating a response. It is covered by membranes called meninges, (inflammation of these membranes is called meningitis) (Fig. 20).

(A)

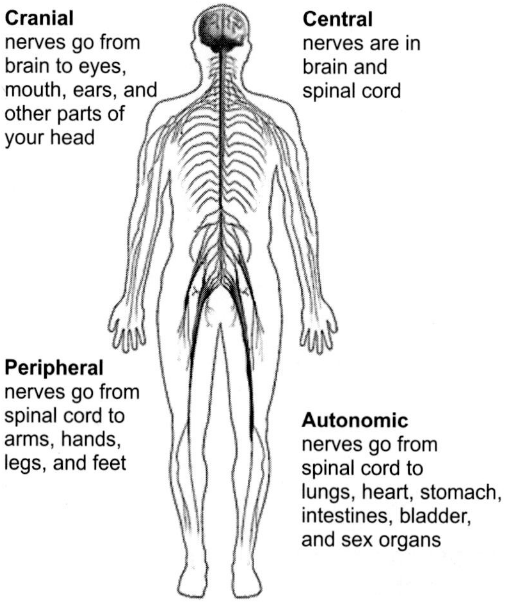

(B)

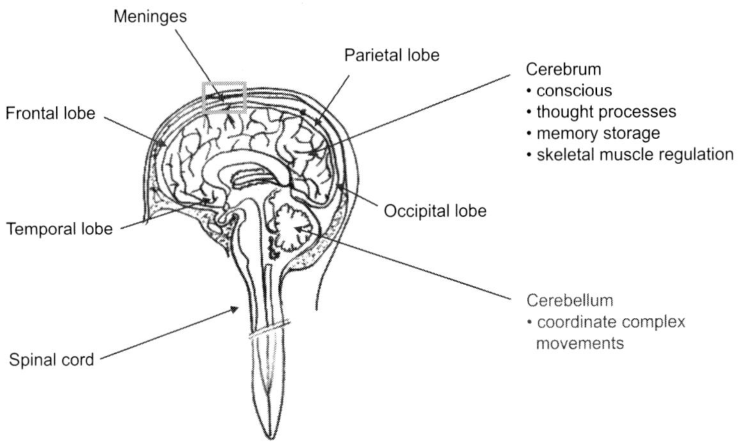


(C)

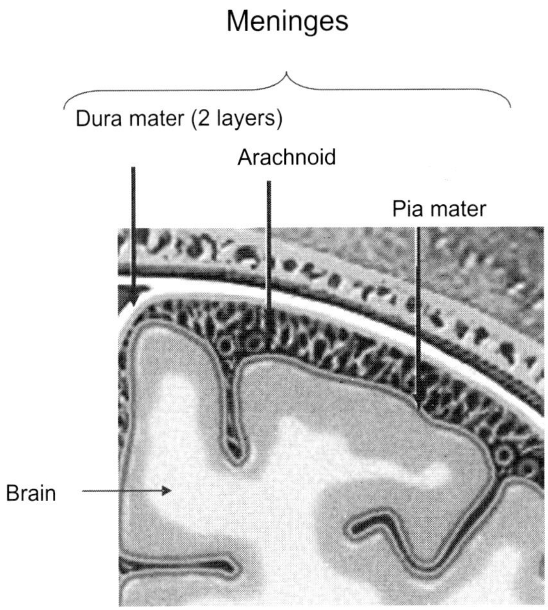


**Fig. 20** (A) Nervous system, (B) Structure of the central nervous system and (C) Meninges, i.e., membranes covering the brain and the spinal cord.
*Colour image of this figure appears in the Colour Plate Section at the end of the Book.*

ii. Peripheral nervous system provides a communication link between the CNS and the rest of the body. It monitors internal and external conditions and reports it to the CNS and then carries the message from the CNS to the body.

## 11.1 Central Nervous System Infections – Viral

### *11.1.1 Viral meningitis (aseptic meningitis)*

It is an inflammation of the membranes covering the brain and the spinal cord. As the CNS is well-protected from the outside environment, it usually develops as a secondary infection (pathogen usually reaches the brain through the bloodstream). The disease is characterized by fever, headache, a stiff neck, confusion, nausea and vomiting. The cerebrospinal fluid (CSF) culture by lumbar puncture does not reveal any agent suggesting aseptic meningitis. There is no specific treatment and the disease usually resolves itself within two weeks but anti-inflammatory agents may be helpful. The causative agents are viruses including enteroviruses (especially Coxsackie and echovirus) which spread by contact with the infected individuals or nasal secretions). Other viruses are mumps that spread as respiratory droplets, herpes simplex virus, varicella (chickenpox), HIV and recently by West Nile virus.

### *11.1.2 Poliomyelitis*

It is damage of the muscles and nerves that can affect the whole body. It may be subclinical and go unnoticed but the disease may develop into the paralytic (associated with fever, headache, a stiff neck and back, sensitivity to touch, constipation, swallowing and breathing difficulty, muscle pain and contraction that progresses to paralysis depending on the location where the spinal cord is affected) and the non-paralytic form (associated with fever, headache, stiffness in the body, muscle tenderness or stiffness, skin rash, vomiting, diarrhoea, fatigue and back pain lasting several days). There is no specific treatment but pain killers and moist heat relieves some of the symptoms. The polio vaccine is effective in preventing the disease. The causative agent is poliovirus that is transmitted by contact with infected individuals or with infected secretions (nasal, mouth, faeces), which multiplies in the gastrointestinal tract and spreads through the blood.

### *11.1.3 Encephalitis and meningoencephalitis*

It is an inflammation of the brain. The disease is transmitted by fever, headache, vomiting, dislike of light, a stiff neck, confusion, drowsiness and may lead to seizures, muscle weakness and paralysis. The virus enters the blood from the primary site of infection and invades the CNS and causes brain tissue swelling (i.e., inflammation known as cerebral edema) that may cause nerve cell damage and bleeding (intracerebral hemorrhage). There is no specific treatment but the use of acyclovir together with anti-seizure drugs, phenytoin and anti-inflammatory drugs, dexamethasone is effective in relieving the symptoms. The causative agents are viruses (such as alphaviruses) transmitted by mosquitoes but rarely coxsackievirus, poliovirus and echovirus, herpes simplex virus, varicella, measles, mumps, rubella, adenovirus, rabies and West Nile virus transmitted through contaminated food, water, contact with infected persons may also cause this disease.

### *11.1.4 Rabies*

It is an acute viral infection affecting the CNS. The disease is characterized by fever, difficulty in swallowing, muscle spasm and loss of feeling in an area of the body, anxiety, stress, drooling, loss of muscle function and convulsions. The symptoms normally appear within several weeks of contracting the virus. The treatment involves the urgent use of passive immunization (an injection of human rabies immunoglobulin). The causative agent is the rabies virus that is transmitted to humans by infected animals (usually dogs, bats) or by exposure of a skin lesion to the saliva of the animal.

## 11.2 Central Nervous System Infections – Bacterial

### *11.2.1 Bacterial meningitis (also called purulent meningitis)*

It is inflammation of the membranes covering the brain and the spinal cord. As the CNS is well-protected from the outside environment, it usually develops as a secondary infection (pathogen usually reaches the brain through the bloodstream). The disease is characterized by fever, chills, headache, severe stiff neck, confusion, dislike of light, nausea, vomiting and decreased consciousness. The treatment requires urgent attention and includes the use of antibiotics. The causative agents are *Neisseria meningitidis* (also called meningococcal meningitis, the antibiotic, benzylpenicillin and chloramphenicol are effective), *Streptococcus pneumoniae* (also called pneumococcal meningitis, the antibiotic, benzylpenicillin and chloramphenicol are effective), *Staphylococcus aureus, Staphylococcal epidermis* (also called staphylococcal meningitis), *Haemophilus influenzae* (also called *H. influenzae* meningitis, the antibiotic, cephalosporin, the recent Hib vaccine is highly effective in preventing this disease), *Listeria monocytogenes* (the antibiotic, benzylpenicillin, tetracycline, amoxycillin, cephalosporins and chloramphenicol are effective), Gram-negative meningitis (caused usually by *E. coli*, *Pseudomonas aeruginosa*, *Enterobacter aerogenes*, *Proteus morganii*, *Klebsiella pneumoniae*) and *Mycobacterium tuberculosis* (also called tuberculous meningitis, treatment with rifampicin, isoniazid, pyrazinamide and ethambutol is effective).

### *11.2.2 Cerebral abscess*

It is a mass of immune cells and pus in the brain following infection of the brain tissue. Pathogens cause infection that results in the inflammation and both immune cells and the pathogens are collected in an area of the brain tissue and become enclosed by membranes forming an abscess. The disease is characterized by a headache, stiff neck and back, fever, chills, weakness, decreased muscle function, vomiting, mental changes such as confusion, drowsiness, seizures and coma. The treatment includes the use of antibiotics, after the correct identification of the organism using needle biopsy and may need surgical removal of the abscess. The causative agents are normally bacteria (predominantly by *Prevotella melaninogenicus*, *Actinomyces* spp. *Haemophilus* spp. *Listeria monocytogenes*, *Fusobacterium* spp.).

## 11.3 Central Nervous System Infections – Fungi

### *11.3.1 Fungal meningitis (also called cryptococcal meningitis)*

It is an inflammation of the membranes covering the brain and the spinal cord. As CNS is well-protected from the outside environment, it usually develops as a secondary infection (pathogen usually reaches the brain through the bloodstream). The disease is characterized by headache, fever, nausea, vomiting, stiff neck, dislike of light, confusion and hallucinations but usually limited to immunocompromised patients. The treatment includes the use of antifungal, amphotericin B or fluconazole. The causative agent is *Cryptococcus neoformans* that is transmitted to humans from the soil.

## 11.4 Central Nervous System Infections – Protozoa

### *11.4.1 Encephalitis and meningoencephalitis*

It is an inflammation of the brain. The disease is characterized by fever, headache, vomiting, dislike of light, a stiff neck, confusion, drowsiness and may lead to seizures, muscle weakness, coma and death. Pathogen enters the blood from the primary site of infection and invades the CNS and causes brain tissue damage. Alternaively, pathogens enter the CNS directly through the nose using the olfactory route. There are no specific treatment and a mixtures of drugs are used for different agents. The causative agents are *Acanthamoeba* spp., *Balamuthia mandrillaris*, *Naegleria fowleri* and *Entamoeba histolytica* transmitted by contact with contaminated objects (swimming in contaminated water) or from the primary site of infection.

# 12. SEXUALLY TRANSMITTED DISEASES

The reproductive system is required to produce and transport male and female gametes for reproduction. Figure 21 shows organs in the male reproductive systems.

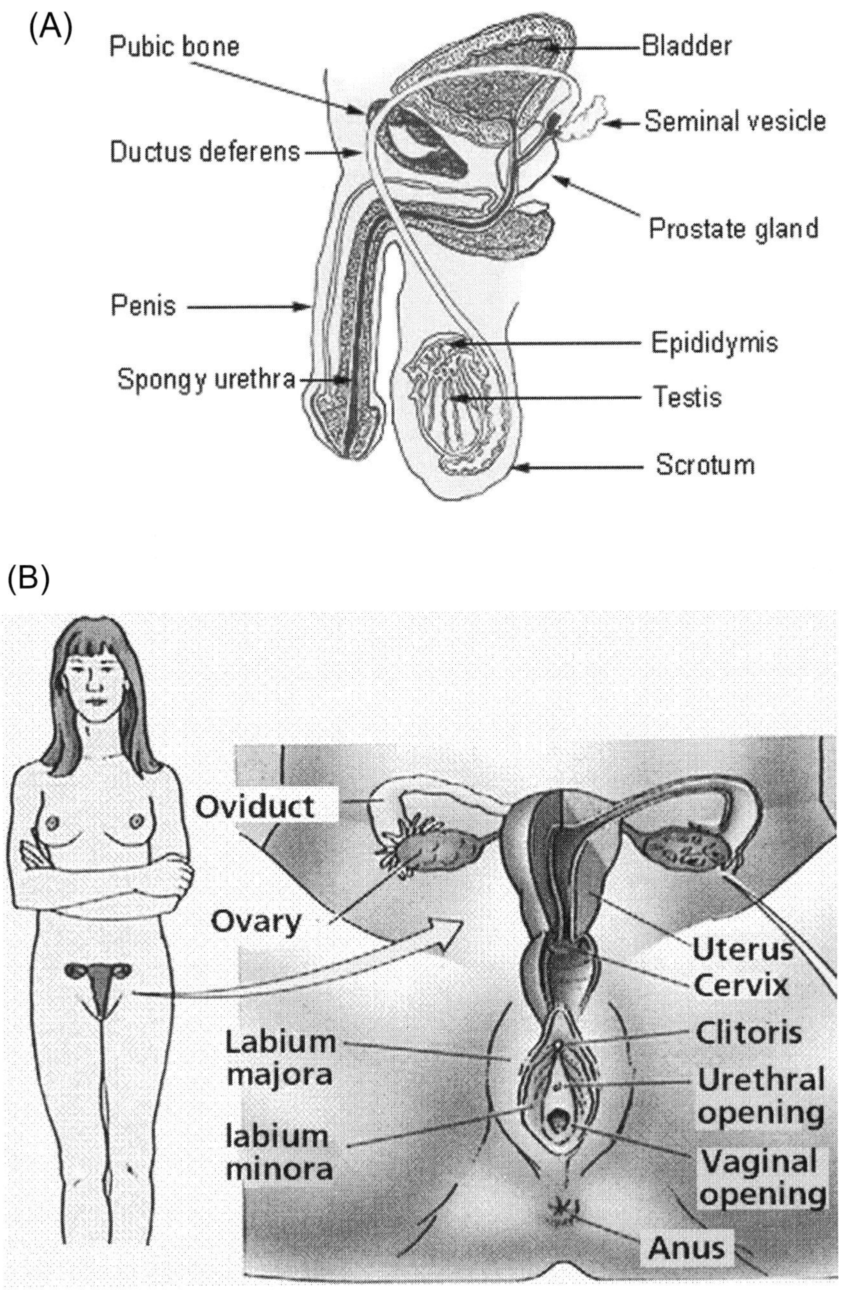


**Fig. 21** Human reproductive system, (A) male and (B) female.
*Colour image of this figure appears in the Colour Plate Section at the end of the Book.*

## 12.1 Sexually Transmitted Diseases – Viral

### *12.1.1 Genital warts*

These are warts grown on genitals (penis, vulva, urethra, vagina, cervix and around the anus). They appear as raised, flesh-coloured lesions with itching of the penis, scrotum, anal or vulval area but the disease may be asymptomatic. Genital warts should be treated to avoid the spread of infection and also due to the fact that HPV has been associated with cervical cancer. The treatment includes cryotherapy or removing them surgically. The causative agent is usually viruse human papilloma virus (HPV).

### *12.1.2 Genital herpes*

It is characterized by repeated eruptions of small, painful blisters on the genitals or on the adjacent skin. It may be asymptomatic but if it develops may exhibit fever, malaise, muscle aches with blisters on the genitals. The treatment includes the use of acyclovir (Zovirax), famciclovir or valacyclovir. The causative agents are herpes simplex virus type 1 (HSV-1, also causes the common cold) and HSV-2.

## 12.2 Sexually Transmitted Diseases – Bacterial

### *12.2.1 Gonorrhea – female*

Usually asymptomatic but if it develops, it can be characterized by vaginal discharge, frequent urination with a burning sensation and painful sexual intercourse. The disease may also spread to through the cervix and uterus into the fallopian tubes (which carry eggs from the ovaries to the uterus) and is called pelvic inflammatory disease. The treatment includes the use of antibiotics, amoxicillin, ceftriaxone, ciprofloxacin or ofloxacin. The causative agent is the bacterium, *Neisseria gonorrhea.*

### *12.2.2 Gonorrhoea – males*

Usually asymptomatic but if it develops, it can be characterized by painful urination and urethral discharge, tender testicles and painful sexual intercourse. The disease may also spread to prostate glands, testicles, and anus resulting in several diseases. The treatment includes the use of antibiotics, amoxicillin, ceftriaxone, ciprofloxacin or ofloxacin. The causative agent is the bacterium, *Neisseria gonorrhea*.

### *12.2.3 Syphilis*

The causative agent is spirochete, *Treponema pallidum*. The disease is characterized by the appearance of a painless sore on the genitals, rectum or mouth called chancre and if untreated it may develop into a rash thoughout the body which involves the palms and soles of the feet, enlargement of the lymph nodes, fever, fatigue and body aches. Infection may recur (within several years) as severe forms, producing destructive lesions of the skin, liver and bones and spread to the CNS. The treatment includes the use of penicillin, ceftriaxone, doxycycline or tetracycline.

#### *12.2.4 Chancroid*

It appears as a genital sore that becomes an ulcer, especially in uncircumcized men. Other symptoms may include enlargement of local lymph nodes, abscess. The treatment includes the use of antibiotics, sulphonamides, ceftriaxone or ciprofloxacin. The causative agent is bacteria.

#### *12.2.5 Granuloma inguinale*

It is the skin damage or granulma formation in the skin as well as the subcutaneous tissue of the genitals. The disease is presented with genital sores, loss of skin colour of genitals or the adjacent skin. The treatment includes the use of antibiotics, tetracycline, doxycycline, ciprofloxacin or erythromycin. The causative agent is *Calymmatobacterium granulomatis*.

### 12.3 Sexually Transmitted Diseases – Fungi

#### *12.3.1 Vaginal yeast infection*

It is characterized by abnormal vaginal discharge and itching, inflammation of vulvar skin, painful urination and intercourse. The treatment includes the use of antifungals, miconazole, nitroimidazole, econazole or clotrimazole. The causative agent is *Candida albicans*.

### 12.4 Sexually Transmitted Diseases – Protozoa

#### *12.4.1 Trichomoniasis*

It is usually a vaginal disease that can also cause damage to the urethra. The disease is characterized by a foul-smell, yellowish vaginal discharge, itching and discomfort during intercourse. Males can also suffer from it with burning after urination. The treatment includes the use of metronidazole. The causative agent is *Trichomonas vaginalis*.

## 13. CARDIOVASCULAR INFECTIONS

Cardiovascular system consists of the heart and extensive network of blood vessels that extend to the rest of the body (Fig. 22). The blood vessels can be divided into two networks, pulmonary circulation where blood completes a round-trip from the heart to the lungs to exchange carbon dioxide for oxygen and systemic circulation where blood completes a round-trip from the heart to the remaining body to deliver oxygen and nutrients to all the tissues.

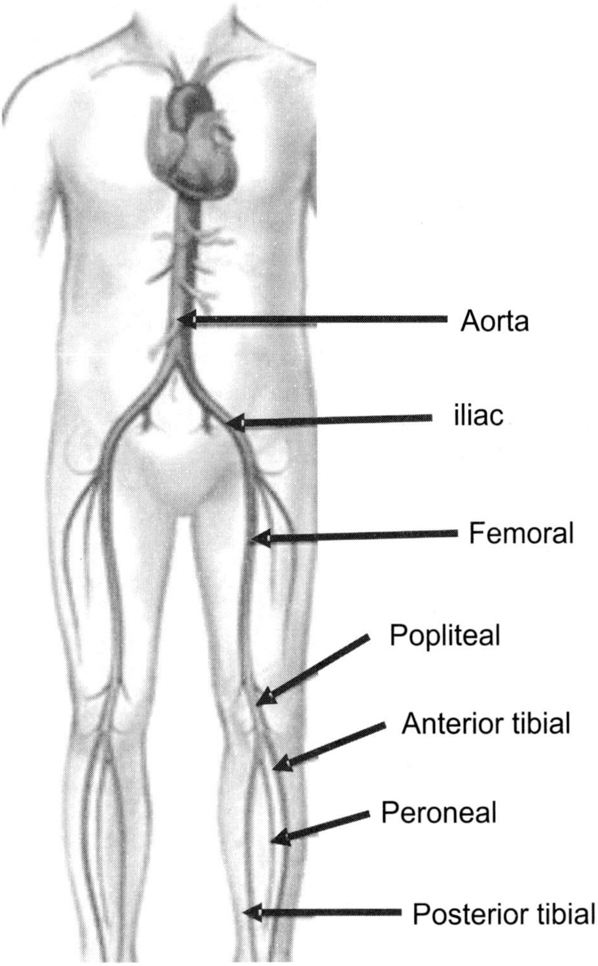


**Fig. 22** Overview of the arterial system in the cardiovascular system.

## 13.1 Pericarditis

It is an inflammation of the pericardium (covering of the heart) and is associated with chest pain, breathing difficulty, a cough, leg and feet swelling, anxiety, fatigue, fever and an enlarged heart with life-threatening consequences. The treatment due to viral infection includes the use of pain killers and anti-inflammatory drugs including ibuprofen, aspirin and corticosteroids. Bacterial pericarditis requires treatment with antibiotics, penicillin, cephalosporin, flucloxacillin together with the above and fungi should be treated with antifungals such as amphotericin together with the above. The causative agents are viruses (usually echoviruses, cocksackie viruses, influenza, or HIV viruses). However, less frequently bacteria (usually *Staphylococcus, Streptococcus pneumococci, Hemophilus influenzae* and *Meningococcus*) and fungal pathogens can also cause this disease.

### 13.2 Myocarditis

It is the inflammation of the heart muscle. The disease may be characterized by fever, chest pain, joint pain, abnormal heart beats, fatigue, breathing difficulty, fever and rash. The treatment due to viral infection includes the use of pain killers and anti-inflammatory drugs including ibuprofen, aspirin and coricosteroids. Bacterial causes require treatment with antibiotics, penicillin, cephalosporin, flucloxacillin together with the above and fungi should be treated with antifungals such as amphotericin together with the above. The causative agents are viruses (usually echoviruses, cocksackie viruses, adenovirus and rarely influenza). However, less frequently bacteria, fungal and protozoal pathogens can also cause this disease.

### 13.3 Endocarditis

It is an inflammation of the inside lining of the heart chambers and heart valves (endocardium). The disease is characterized by fatigue, weakness, fever, chills, muscle aches, breathing difficulty, blood in urine, swelling of the feet and legs and joint pain. The treatment involves the use of antibiotics, benzylpenicillin, doxycycline, tetracycline, amoxycillin or erythromycin together with gentamicin. The common causes are viruses, bacteria (usually *Streptococcus viridans* and *Staphylococcus and Entercoccus*), protozoa and rarely caused by fungi.

## 14. BONE AND JOINT INFECTIONS (SKELETAL SYSTEM)

It is part of the skeletal system, which includes bones and the cartilage that provides the structural support for the body. It stores minerals such as calcium, phosphate, fats and produce red blood cells and other blood elements.

### 14.1 Osteomyelitis

It is an acute or chronic infection of the bones. It is usually a secondary infection that spreads to the bones via the bloodstream and produces pus, forming an abscess in the bone and limiting its blood supply and may cause permanent damage to the bone. The disease is characterized by pain, local swelling, fever, nausea and malaise. The treatment includes the use of antibiotics, clindamycin, flucloxacillin or cefuroxime but the causative agent should be identified for the correct antibiotic treatment using needle biopsy and may also require surgical removal of the bone in chronic infection. The causative agent is bacteria (*Haemophilus influenzae*, *Staphylococcus aureus*, *Streptococcus pyogenes*).

### 14.2 Malignant Otitis Externa

It is an inflammation and damage of the bones and the cartilage at the base of the skull. The disease spreads from the ear infection and is characterized by yellowish pus from the ear, hearing loss, ear pain, itching and fever. Due to the difficulty of getting to the bacteria in the bones, treatment requires the prolonged use of antibiotics, neomycin, framycetin or clioquinol. If infection is due to fungus, clotrimazole is effective. The causative agents are usually *Psedomonas* spp. and very rarely by fungus, *Aspergillus*.

### 14.3 Arthritis

It is an inflammation of one or more joints due to the breakdown of the cartilage (the cartilage normally protects the bones and allows smooth movement) (Fig. 23). The damage or lack of cartilage causes the bones to rub together causing pain and swelling. The disease is characterized by pain, swelling, stiffness, redness of the skin and reduced ability to move. The treatment includes the use of antibiotics or antifungals (not for non-infectious diseases) and anti-inflammatory agents (ibuprofen), together with the use of glucosamine and chondroitin (building blocks of the cartilage), vitamin E, a healthy diet, use of capsaicin cream (derived from hot chili peppers) over the skin of the affected joints and exercise to maintain healthy joints. In severe cases, the use of acetaminophen, ibuprofen, corticosteroids and cyclo-oxygenase-2 inhibitors inhibiting inflammation. If infectious, the causative agent is bacteria or viruses (or due to non-infectious disease, such as general wear out and autoimmune diseases).

## 15. DISSEMINATED INFECTIONS

Given the environmental conditions and host susceptibilty, the majority of microbial pathogens have the abiltiy to disseminate through the blood-stream from the primary site of entry into the body to infect multiple organs, leading to their dysfunction and possibly death. Immunocompromised patients (with a weakened immune system) are particularly susceptible to

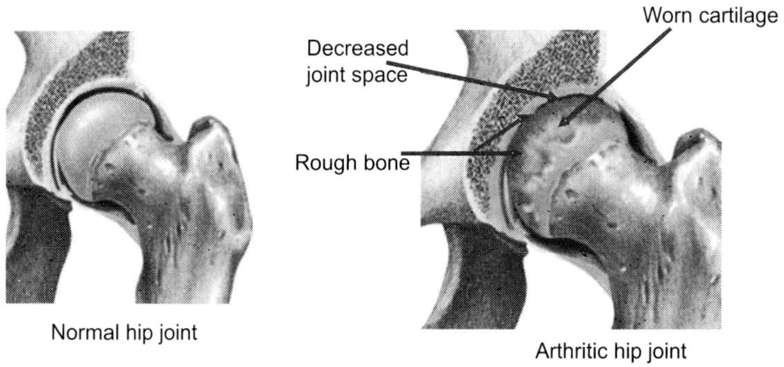


**Fig. 23** Breakdown of cartilage in arthritis (source: www.nlm.nih.gov/medlineplus). *Colour image of this figure appears in the Colour Plate Section at the end of the Book.*

disseminated infections. Disseminated infections can be caused by viruses, fungi, protozoa but predominantly by bacteria, e.g., septeicemia or toxic shock syndrome.

## 15.1 Sepsis and Septicemia

It is an overwhelming infection of the bloodstream by toxin-producing bacteria and can involve various organs. The disease is characterized by fever, hyperventilation, chills, skin rash, rapid heart beat, confusion, changes in mental status, shock and death. If bacteria are also present in the blood (bacteremia), then it is known as septicemia. It is an overwhelming infection which results in low blood pressure (hypotension) resulting in shock (dysfunction of heart, liver, kidneys, brain and other organs). The treatment needs to be urgent identifcation of the causative agent followed by appropriate antibiotics as well as oxygen therapy in the Intensive Care Unit (ICU). The causative agent is usually bacteria.

## 15.2 Toxic Shock Syndrome

It is characterized by fever, rash, hypotension and may involve different organs (including cardiovascular, renal, skin, mucosa, gastrointestinal, musculoskeletal, hepatic, hematologic, and the CNS). Skin infection appears as rash appearing on the trunk, spreading to the arms and legs, and involving the palms and soles with desquamation (skin peel off) and associated with high fever, hypotension. The involvement of other organs is associated with acute respiratory distress, necrotizing fasciitis, altered consciousness and mucosal inflammation (such as vaginitis, conjunctivitis, pharyngitis). The use of anti-staphylococcal antibiotics including nafcillin (a penicillinase-resistant synthetic penicillin), flucloxacillin or cloxacillin, while cephalosporins, rifampicin or trimethoprim are also effective. In addition, aggressive fluid resuscitation is needed as well as oxygen therapy and hemodynamic support. It is caused by the Gram-positive bacteria (usually *Staphylococcus aureus* and *Streptococcus pyogenes*).

# 16. OTHER IMPORTANT INFECTIONS

## 16.1 Tuberculosis

The disease is characterized by the development of granulomas in the infected tissue (often takes years to develop) and is associated with fatigue, cough possibly with blood, fever, chest pain, breathing difficulty and is known as pulmonary tuberculosis (Fig. 24). If it results in the accumulation of fluid in the lungs it is called pleural infusion tuberculosis. In a minority of patients, it may invade into the bloodstream and affect other organs and is called disseminated tuberculosis and may also cause arthritis. The treatment includes the use of antitubercular drugs, rifampicin, isoniazid, pyrazinamide and ethambutol for several months or longer. The causative agent is *Mycobacterium tuberculosis* which is spread by respiratory droplets from the cough and sneezing of infected individuals.

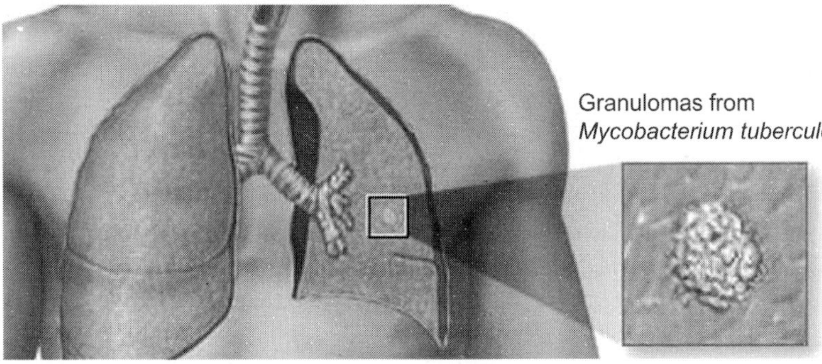


**Fig. 24** *Mycobacterium tuberculosis*-infected lungs.

## 16.2 Yellow fever

It is characterized by high fever, chills, headache, muscle aches, vomiting, and backache. After a brief recovery period, the infection can lead to kidney failure, liver failure and shock. Liver failure causes jaundice (yellowing of the skin and the whites of the eyes), hence the disease is called yellow fever. There is no specific treatment but the use of ribavirin together with symptomatic therapy may be effective in relieving the symptoms. The vaccine is the best option. People travelling to endemic areas should be vaccinated (the vaccine is available). The causative agent is yellow fever virus that is transmitted to humans by infected mosquitoes (*Aedes* spp.), which can carry the virus from human to human or monkey to human.

## 16.3 Dengue Fever

It is fever characterized by a rash, muscle and joint pains, headaches, nausea, vomiting and enlarged lymph nodes and may rarely lead to hemorrhage fever leading to bleeding from the nose, mouth. The treatment includes rehydration together with targetting the fever with acetaminophen. The causative agent is the dengue virus which is transmitted to humans through mosquito bites (*Aedes* spp.).

## 16.4 Malaria

It is characterized by fever, chills and anemia. The causative agent is *Plasmodium* spp. which is transmitted to humans through mosquito bites. The disease is characterized by alternate periods of chills, fever, sweating, headache, nausea, anemia, jaundice and may also lead to death. The treatment includes the use of mefloquine, chloroquine, hydroxychloroquine or malarone.

## 16.5 Cutaneous Leishmaniasis

It is a skin infection causing skin lesions or ulcers. The disease is characterized by macule or papule, skin ulcers that heal slowly and may become dark, dry skin. It can also cause a systemic infection called visceral leishmaniasis which is associated with diarrhoea, fever and a cough, fatigue, weight loss, and scaly skin. The treatment includes the use of meglumine antimonate, sodium stibogluconate, pentamidine or amphoterocin B. The causative agent is *Leishmania* which is transmitted to humans via the bite of the sandfly.

## 16.6 Tetanus (also called Lockjaw)

It is characterized by tightening of the jaw muscle and is associated with stiffness of muscles (the neck, chest, abdomin and the back), muscular seizures, fever, sweating, hand or foot spasm and drooling. The treatment includes the use of antitoxin (tetanus immunoglobulins) as well as penicillin, clindamycin or erythromycin to kill the bacterium. The disease is caused by a toxin produced by the bacterium, *Clostridium tetani* which is found in the soil and introduced into the human body through skin lesions.

## 16.7 Typhoid Fever

It is characterized by rash, diarrhoea possibly with blood, fever, malaise, abdominal pain, fatigue, severe confusion and hallucinations. Infection may spread to various organs. The treatment includes excessive intake of fluids and antibiotics, ciprofloxacin or oxytetracycline. The causative agent is *Salmonella typhi* which are transmitted via contaminated food or water.

## 16.8 Cholera

This is an infection of the small intestine. The disease is characterized by massive watery diarrhoea, dry skin and mucosal membranes, tiredness, abdominal cramps, nausea, low urine and dehydration. The treatment includes excessive intake of fluids as well as application of antibiotics, tetracycline. The causative agent is *Vibrio cholerae* which is transmitted via contaminated water, food.

## 16.9 Plague

Plague may be (i) the infection of the lymph nodes (known as bubonic plague, associated with high fever, swelling of lymph glands, discomfort, muscular pain, severe headache, seizures), (ii) infection of the lungs (called pneumonic plague associated with a severe cough, bloody sputum, breathing difficulty), (iii) infection of the blood (called septicemic plague associated with fever, nausea, vomiting, low blood pressure, blood clots, dysfunction of several organs). Infected humans can spread it to others via respiratory droplets. The treatment requires immediate attention and includes streptomycin, chloramphenicol or tetracycline. The causative agent is *Yersinia pestis* which is transmitted to humans by contact with rodents (e.g., rats) or their faeces.

## 16.10 Lassa Fever

Symptoms of Lassa fever appear within several days of contracting the disease and include fever, chest pain, a sore throat, back pain, a cough, abdominal pain, vomiting, diarrhoea, conjunctivitis, facial swelling and mucosal bleeding. In addition, it may involve neurological problems such as hearing loss, tremors and encephalitis. Due to the varied symptoms, clinical diagnosis is difficult. The treatment includes supportive care together with the use of antiviral, ribavirin, especially during the early course of infection. The causative agent is Lassa fever virus, which is transmitted to humans by contact with rodents or their faeces.

## 16.11 Anthrax

It is infection of the lungs, skin or gastrointestinal tract. If infection initiates on the skin due to lesion, it is called cutaneous anthrax (associated with blisters and extensive ulcers, swollen lymph nodes, headache and malaise) and if they are inhaled into the lungs, it is called inhalational anthrax (associated with fever, malaise, headache, a cough, shortness of breath and possibly shock). Feeding of bacterium will result in the gastrointestinal anthrax (nausea, vomiting, possibly with blood) and bloody diarrhoea. The treatment includes early diagnosis together with the application of antibiotics, penicillin, doxycycline and ciprofloxacin. The causative agent is *Bacillus anthracis* which is spread from animals (sheep, goats) to humans or from the soil.

## 16.12 Tularaemia

It is characterized by enlarged lymph nodes, headache, muscle pains, shortness of breath, fever, chills, sweating and weight loss. The treatment includes the use of streptomycin and tetracycline. The causative agent is *Francisella tularensis* which is transmitted to humans through contact with animal tissues (ingested infected meat or through skin lesions) or ticks.

## 16.13 Dengue Haemorrhagic Fever

It occurs when some one with immunity to the dengue virus becomes infected with a different type. It is a fever characterized by rash, muscle and joint pains, headaches, nausea, vomiting, an enlarged liver and a shock-like syndrome leading to death. The treatment includes rehydration together with targetting symptoms, acetaminophen for fever and oxygen therapy. The causative agent is viruses (arboviruses) which are transmitted to humans through mosquito bites.

## 16.14 Ebola Haemorragic Fever

It is characterized by fever, a headache, rash, fatigue, malaise, backache, vomiting, bleeding abnormalities, shock and death (up to 90% mortality). The symptoms appear within one week of contracting the virus. The causative agent is the Ebola virus which is transmitted to humans via contact with infected animals or their products. There is no treatment.

# 17. HUMAN DEFENCE MECHANISMS

Once bacterial pathogens gain access into the human body, they are encountered by a highly professional defence system. Traditionally, the human defence system has been divided into two components:

1. Non-specific / constitutive / innate immune response that include: skin which acts as a physical barrier, neutrophils that police the bloodstream and attack any foreign invaders, complement and cytokines that direct the activities of neutrophils, and natural killer cells which kill the virus-infected cells.
2. Specific / inducible / acquired immune response that include: antibodies (produced by B-lymphocytes) and T-lymphocytes.

These defence systems do not operate independently but communicate with each other to build an effective defence against the invading organism.

## 17.1 Non-specific Immune Responses

### *17.1.1 Defence of surfaces*

The outer layers of the skin are dry, thick layers of keratinized cells which provide a physical barrier to pathogens. Natural openings in the skin such as pores, hair follicles, and sweat glands that are protected by secretion of toxic chemicals such as fatty acids and lysozyme which combine to prevent colonization by potential pathogens. Generally, pathogens can only breach this barrier through wounds. However, underneath the skin is skin-associated lymphoid tissue (SALT). In SALT, the resident Langerhans cells (similar to macrophages) attack microbial pathogens and initiate specific immune defences.

Inside the body, the intestinal tract, respiratory tract, vaginal tract and bladder are covered in mucosal membranes. The cells in these membranes are protected by a thick layer of mucus/mucin. The mucin consists of proteins and polysaccharides which prevent microbes from reaching the epithelial cells. Mucin also contains:

1. Lactoferrin – iron binding protein that deprives microbes of iron.
2. Lysozyme – an enzyme that digests cell walls of bacteria, by disrupting the N-acetyl muramic acid – N-acetyl glucosamine linkages. It is present in tears, saliva, sweat, and other mucosal secretions.
3. Defensins – small proteins that make holes in the microbial membranes by a charge or voltage-dependent mechanism.

Mucin is constantly shed and replaced, together with any trapped microbes. In addition, low pH (2) in the stomach, cilia movement over trachea, flushing of urinary tract and competition from the normal flora prevents colonization of these by pathogenic microbes. In addition, infection may be limited by the short life of intestinal epithelial cells, which are shed approximately every 30 h, together with any microbial pathogen. Similar to the skin, mucoal membranes are associated with the specific immune defence by mucosa-associated lymphoid tissue (MALT). For intestinal tract, it is known as gastrointestinal-associated lymphoid tissue (GALT). In GALT, the resident macrophages ( with a similar function to Langerhans cells of SALT) initiate specific defences.

In addition to the above, the skin and mucus membranes of the body are home to a variety of virus, bacteria, protozoa and fungi. This normal flora plays an important role in protecting the body by competing with potential pathogens, a situation known as microbial antagonism. The normal flora achieves this by: (i) consuming the available nutrients, making them unavailable to pathogens; (ii) stimulating the body's defences; and (iii) occupying binding sites required by pathogens.

The violation of initial defences may lead to inflammation. Inflammation is tissue reaction to infections or injury characterized by redness (increased blood flow), swelling (increased extravascular fluid and phagocyte infiltration), heat (increased blood flow and pyrogens, fever-inducing agents), pain (local tissue destruction and irritation of sensory nerve) and loss of function. Inflammation leads to recruitment of neutrophils to the infected site, which leads to ingestion and destruction of microbes (Fig. 25). Neutrophils are also known as polymorphonuclear leukocytes (PMNs) indicating that the nucleus is a multi-lobed structure that may appear to be multi-nuclei when viewed in cross-sections. The PMNs are primarily found in the bloodstream. The steps of ingestion and killing by phagocytic cells involve the following:

1. Attachment of the microbe – formation of pseudopodia.
2. Uptake of the microbe in a vacuole 'phagosome'.
3. Fusion of lysosome (containing antimicrobial chemicals and enzymes) with phagosome. Lysosome contain myeloperoxidase enzyme which is normally inactive but following fusion with phagosome, this enzyme becomes activated and produces superoxide and hypochlorite. Superoxide is a toxic form of oxygen that can oxidize and inactivate proteins and other molecules on microbial surface. When mixed with chlorine, it forms hypochlorus acid, highly toxic to invading organisms. In addition, toxic nitric oxide (NO) is produced in response to invading microbes.
4. The overall result is killing and digestion of the ingested microbe. Often death of the PMN also occurs and if in large numbers, the dead PMNs form pus.
5. Release of digested products to the outside

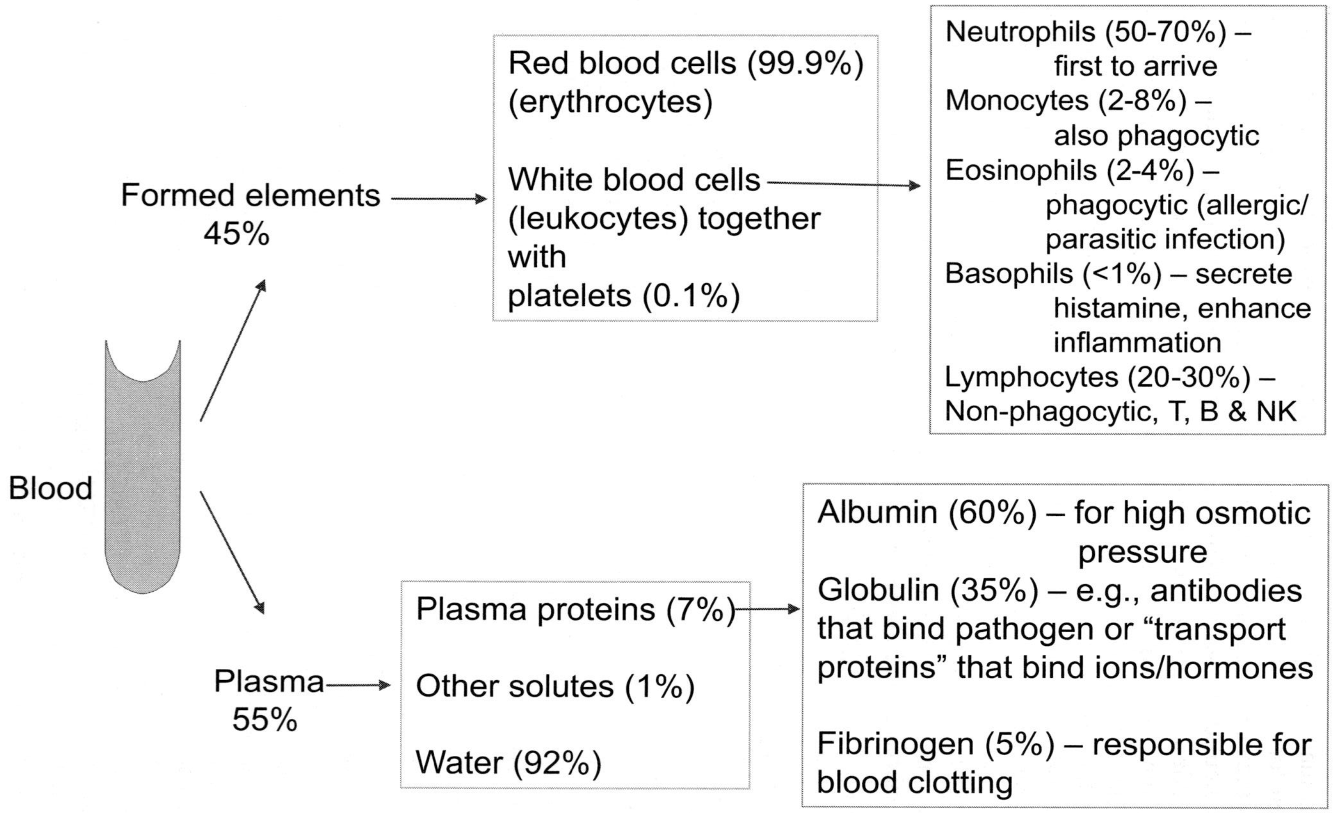


**Fig. 25** Components of blood.

Other phagocytic cells are initially monocytes, which leave the blood and mature into phagocytic macrophages. Macrophages are named for their location. For example, wandering macrophages leave the blood by squeezing through the cells, lining the capillaries and target microbes, as for neutrophils/eosinophils. Other macrophages do not wander. These include alveolar macrophages of the lungs, microglia of the central nervous system and Kupffer cells of the liver. Generally, fixed macrophages phagocytoze microbes within a specific organ. Another group of phagocytes are dendritic cells, which are multibranched and are scattered throughout the body, particularly the skin and mucous membranes, where they await microbial invaders. In addition, plasma (a cellular part of the blood) also contains: (i) lactoferrin, an iron binding protein that deprives pathogens of iron; (ii) lysozyme which digests cell walls of pathogens; and (iii) defensins, that interfere with pathogen intracellular signalling pathways and metabolism, and may produce holes in the pathogen membranes.

### *17.1.2 Complement*

In addition to neutrophils, blood/plasma contains a potent complement system. A complement is a set of proteins produced in part by the liver which circulate in the blood and are activated sequentially by proteolytic cleavage, a process called complement activation. The complement attaches itself to the invading microbe and directly causes damage to the microbial cell through a process known as membrane attack complexe (MAC) or helps phagocytes (macrophages and neutrophils) to eliminate the microbe from the body (Fig. 26). There are nine complement proteins, C1 – C9. There are two pathways of complement activation which is as follows:

1. The classical pathway which is initiated by antibody-antigen complexes and
2. The alternative pathway which is triggered by microbial surface molecules such as cell membranes.

The activation causes complement proteins to be proteolytically cleaved. The cleaved products are indicated by small fragment 'a' or large fragment 'b'. In the classical pathway, antibodies specifically bind to the microbial antigens. A complement protein, C1, binds to the antibody-antigen complex and becomes an active enzyme. Active C1 splits several molecules of C2 and C4. The fragments of C2 and C4 combine to form an enzyme, C4bC2b (C3 convertase), that splits C3 into C3a and C3b. C3b combines with the remaining fragments of C2 and C4 to form an enzyme that cleaves C5 into C5a and C5b. C5b combines with C6, C7, C8 and several molecules of C9 to form a membrane attack complex (MAC). A MAC drills a circular hole in the pathogen's membrane leading to hypotonic lysis of the cell. In contrast, the alternate pathway occurs independently of antibodies. It begins with

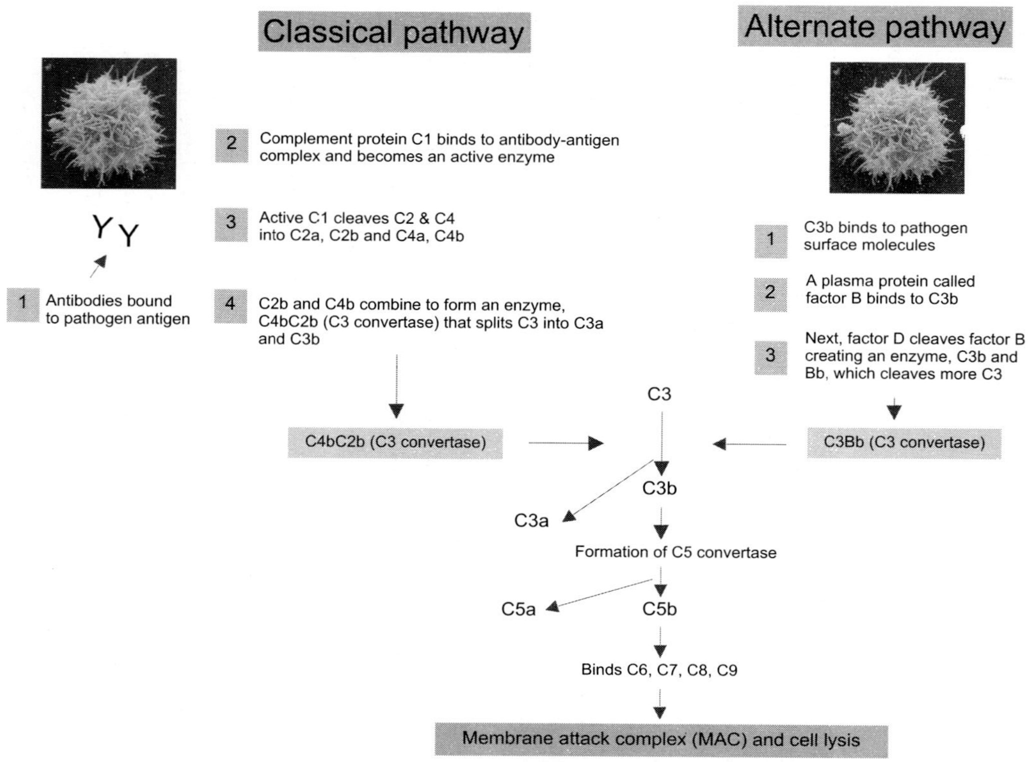


**Fig. 26** Two pathways by which complement is activated, resulting in the formation of the membrane attack complex (MAC) in the host cell membranes.

cleavage of C3 into C3a and C3b. This occurs naturally in the plasma but at a slow rate and C3b is further degraded. However, in the presence of a microbe, C3b binds to the microbial surface, to a protein called factor B. Another plasma protein, factor D, cleaves factor B creating an enzyme composed of C3b and Bb, which continues with the complement pathway and MAC. The ultimate effect of complement is damage of the microbe by creating pores (i.e., MAC) or their opsonization (pathogen uptake by phagocytes).

### *17.1.3 Cytokines*

Cytokines are large number of proteins that are usually less than 20 kDa which serve as a hormone-like function in allowing cells to communicate with each other. However, hormones act in an endocrine manner, i.e., produced in one organ and travel through the bloodstream to another organ. Instead, cytokines are produced and act locally in a paracrine and autocrine manner. Thus, the role of cytokines is to enable cells to communicate with each other locally. They are produced primarily by endothelial cells (which form blood vessels) and phagocytes (macrophages and neutrophils). There are many cytokines which are divided into families (Table 1). Cytokine producing cells express CD14 receptor on their surface, which binds to the specific bacterial components such as lipopolysaccharide (LPS). This leads to the release of cytokines. The function of cytokines is like smoke-signals, as they attract phagocytes to the site of

**Table 1** Examples of major cytokine families.

| *Family* | *Member (examples)* | *Comments* |
|---|---|---|
| Interleukin (IL) | IL-1 to IL-32 | Secreting by different cells and function differently (e.g., IL-1 is involved in fever, inflammation, T-cell activation, macrophage activation; IL-2 is involved in proliferation, etc.). |
| Interferon (IFN) | IFN-α, | Leukocyte IFN (inhibit viral replication) |
| | IFN-β, | Fibroblast IFN (inhibit viral replication) |
| | IFN-γ | Lymphocyte IFN (many functions including macrophage activation, induction of MHC-I & II) |
| Tumor necrosis factor (TNF) | TNF-α | Secreted by monocytes and other cells (fever, activates macrophages and endothelial cells) |
| | TNF-β | Secreted by T-cells (activity similar to TNF-α) |
| Colony stimulating factor (CSF) | G-CSF, M-CSF, GM-CSF, | Secreted by leukocytes, endothelial cells (helps differentiate bone marrow cells into particular cell type, e.g., neutrophils, macrophages) |
| Chemokine | MCP, Eotaxin | Controls the migration of cells between and within tissues |
| Growth factor | TGF, IGF | Anti-inflammatory effects |

infection and link non-specific and specific immune systems. The major cytokines are interleukins, colony-stimulating factors, interferons, tumour-necrosis factors, chemokines and growth factors. Cytokines control many aspects of cell behaviour both in normal physiology as well as play a crucial role in the infection. Figure 27 provides a summary of innate / non-specific immune responses.

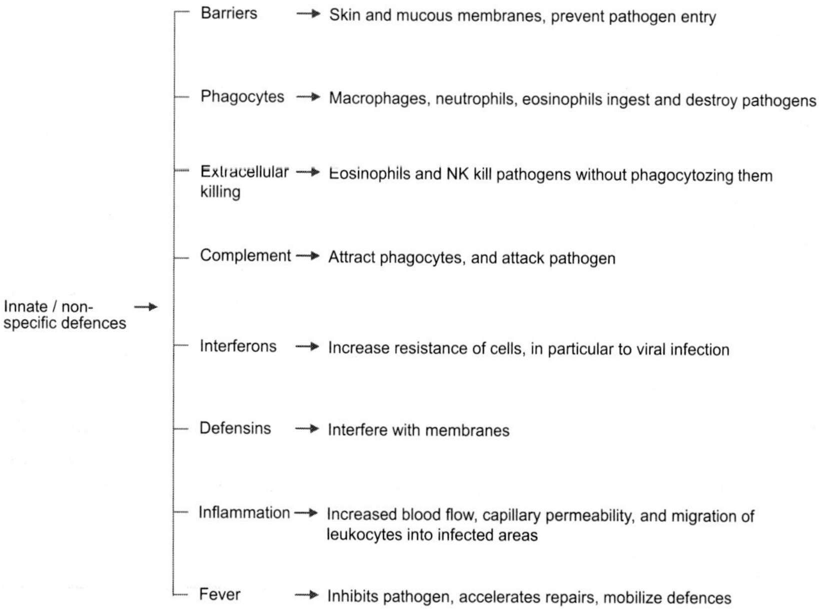


**Fig. 27** Summary of components of the innate / non-specific defences.

## 17.2 Specific Immune Response

This system involves cells that respond to the invading pathogen in a specific fashion. These include cytotoxic T-cells (CD8 cells), T helper cells (CD4 cells), B cells (produce antibodies) and active macrophages (Table 2). Macrophages are phagocytes similar to neutrophils which are mostly found in the tissues. Macrophages are first produced by stem cells as monocytes, which circulate in the bloodstream. On attraction to a site of action or entry into a tissue they differentiate into macrophages. The invading pathogens are taken up by macrophages and killed (like in neutrophils) but instead of releasing the digested products to the outside, some microbial products are loaded onto a complex of proteins called the major histocompatibility complex class II proteins (MHC II) and presented on the cell surface. Hence, these cells are called antigen-presenting cells (APCs). Why certain microbial

**Table 2** Cells involved in immune defence against microbes.

| | |
|---|---|
| Eosinphils and neutrophils (PMN) | Phagocytic cells |
| Basophils | Non-phagocytic cells, that release inflammatory chemicals |
| Monocyte | Immature macrophage – limited killing but produce cytokine |
| Macrophage | Phagocytic cell that kills, and present antigens to other immune cells |
| Dendritic cell | Phagocytic cell, active in presenting antigens to other immune cell |
| NK cell | Kills cells infected with cytoplasmic pathogens |
| B cells | When activated, produce antibodies |
| CD8 T cells | Cytotoxic T cells, kills cells containing cytosolic bacteria |
| CD4 T cells | Helper cells for B cells and activates macrophages |
| Fibroblasts | Act in repair after infection |
| Platelets | Produce clotting, which act as barrier to pathogen dissemination and blood loss |

products are selected to be expressed on APCs (macrophages, dendritic, B cells) and not other microbial products is not known. Of interest, macrophages in the brain are called microglial cells, while macrophages in the liver are called Kupffer cells. The B cells also act like APCs in taking up and processing foreign antigens and loading them onto the MHC II complex. Finally, the antigen-MHC II complex is recognized by T helper cells.

1. In the case of T helper-macrophage interactions, T helper cells become activated and produce cytokines, such as interferon-gamma (IFN-$\gamma$) and interleukin-2 (IL-2), which activate macrophages and stimulate B cells to produce antibodies.
2. In the case of T helper-B cell interactions, T helper cells become activated as above, which in turn stimulates B cells to differentiate into antibody-producing cells, that is, plasma cells.

This approach is used in response to extracellular microbial pathogens. In contrast, if the microbial pathogen is intracellular, their products are loaded onto the MHC I complex. The MHC I displays proteins on almost all cells. Finally, the antigen-MHC I complex is recognized by cytotoxic T-cells. Cytotoxic T-cells kill the host cells displaying antigen-MHC I complex, thus killing any intracellular microbe. This is particularly useful against viral infections.

### *17.2.1 Antibodies*

Antibodies are proteins called immunoglobulins (Igs). They are made of two heavy and two light chains, which create two functionally important areas (Fig. 28). The region that recognizes and binds the antigen is called the Fab region, while the Fc region binds to phagocytes (phagocytes have an Fc receptor). The result of an antibody binding to a specific antigen is:

- Enhanced phagocytosis by providing a binding site (Fc) for phagocytes
- Microbes trapped in mucin
- Complement activation
- Neutralization of toxins
- Inactivation of proteins
- Inhibition of binding of toxins/microbes to the host cells.

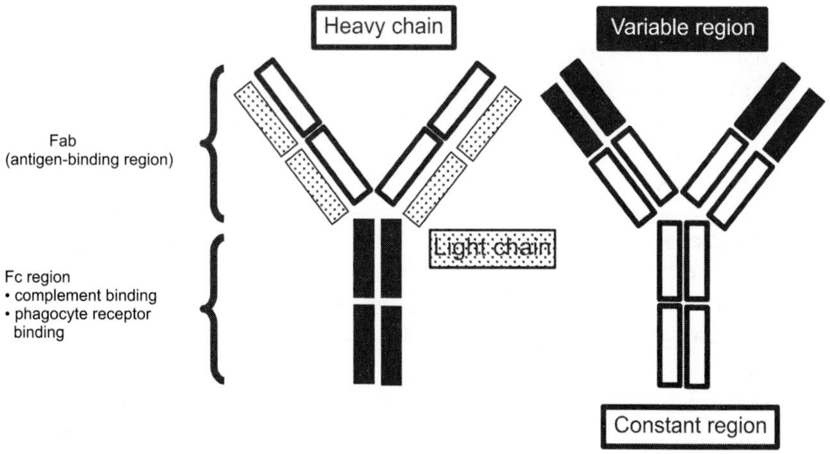


**Fig. 28** Structure of an antibody.

This ultimately results in the neutralization and removal of microbes and/or their toxins from the host. The antibodies present in the serum are IgG (70–75% of total antibodies) and IgM (10% of total antibodies), while IgA (15% of total antibodies) are present in secretions such as milk, tears, saliva and mucosal secretions, and are called secretory IgA (sIgA). In addition, IgE and IgD are present in the serum and are involved in allergic reactions.

## 18. ANTIMICROBIAL

### 18.1 Antibiotics

Molecules that inhibit the growth of, or kill bacteria are called antibacterial agents. These compounds are normally isolated from microbes and are called antibiotics. There are many millions of antibiotics, produced mainly by soil bacteria and fungi which are active against bacteria but only a few can be used to control or treat human bacterial diseases. These are few which although are toxic to bacteria, but have no significant toxic effects on the human host. The reason for this so called selective toxicity is that the site at which these antibiotics act are either unique to bacteria such as peptidoglycans or very different between prokaryotes and eukaryotes, such a ribosomes and nucleic acid synthesis. Some of the factors which affect the antibiotic therapy are listed below:

- The antibiotic must reach the site of infection in the host.
- The antibiotic has to reach its target site. This is easier for antibiotics such as penicillin which act on peptidoglycan than for those like tetracycline which must penetrate through the plasma membrane to reach their target sites, the ribosomes.
- Gram-negative bacteria are often intrinsically resistant to the action of antibiotics due to the presence of the outer membrane which acts as an additional barrier for the antibiotic to cross and protects the peptidoglycan.
- Broad-spectrum antibiotics are effective against a wide range of different Gram-positive bacteria whereas other antibiotics may have only a narrow range.
- All pathogenic bacteria must be eradicated from the host by either inhibiting the growth of the microbe (bacteriostatic antibiotics), which can then be removed by the immune system or by killing them directly (bactericidal antibiotics).

Antibiotics have proved to be of great benefit to human kind and have ensured that people no longer need to die from diseases such as wound infections. However, recent studies have clearly shown that bacteria can become resistant to the action of antibiotics. The mechanisms by which they do this include: the production of enzymes that break down the antibiotic, reduce the permeability to the antibiotic, and alterations to the target site. Antibiotic resistance may arise by mutation but more often the genes for antibiotic resistance are transferred between bacteria by conjugation, transduction and transformation. Antibiotic resistance carried by plasmids has caused particular concern as these plasmids may carry genes that confer resistance to many different antibiotics at the same time. Multiple-resistant bacteria are therefore becoming a problem particularly in the hospital environment and the fear is that it will not be long before there is a bacterial strain which is untreatable by all known antibiotics. Thus there is a clear need to identify novel targets in bacterial pathogens as well as seek alternative approaches to develop preventative and therapeutic approaches.

# 19. COMMONLY USED ANTIMICROBIALS

## 19.1 Antibacterial Agents

**Beta-lactam antibiotics**

These include

1. Penicillin – which can be divided into the following major groups based on the activity.
   1.1. Benzylpenicillin (also known as penicillin G) – derived from *Penicillium chrysogenum* and effective against Gram-positive and some Gram-negative bacteria.
   1.2. oxazolyl penicillins (penicillinase-resistant penicillins) such as cloxacillin, oxacillin, flucloxacillin – developed to overcome the penicillin resistance.
   1.3. Aminopenicillin including ampicillin and amoxicillin – have broad spectrum activity.
   1.4. Penicillin V (phenoxymethylpenicillin) is particularly stable to gastric acid and is frequently used via the oral route but are less effective than others.
2. Cephalosporins including cefazolin, cefamandole, cefuroxime, cefazidime ceftriaxone, cefixime, cefotaxime – these are naturally occurring compounds, which are structurally similar to penicillin and exhibit activity against Gram-positive and Gram-negative bacteria.
3. Aminoglycosides – the first agent was streptomycin isolated from *Streptomyces* spp. and others have been developed including kanamycin, gentamicin, tobramicin, netilmicin and inhibit protein synthesis at the translational level.
4. Tetracycline including doxycycline – first isolated from *Streptomyces* spp. and are active against Gram-positive, Gram-negative and spirochetes and *Ricketssia*, by inhibiting protein synthesis at the translational-level.
5. Chloramphenicol – first isolated from *Streptomyces* spp. and inhibit protein synthesis and has broad-spectrum activity as in tetracycline.
6. Quinolones including nalidixic acid and Fluoroquinolones such as ciprofloxacin, ofloxacin inhibit the transcriptional process active against Gram-negative and some Gram-positive bacteria.
7. Vancomycin – It is a glycopeptides antibiotics first isolated from *Streptomyces* spp. that target Gram-positive bacteria by inhibiting peptidoglycan synthesis (cannot penetrate the outer membrane of Gram-negative bacteria and is not effective against them).
8. Teicoplanin – It also is a glycopeptides antibiotic similar to vancomycin but isolated from *Actinoplanus teichomyceticus*.

9. Metronidazole – It is a nitroimidazole drug effective against anaerobic bacteria and protozoa, drug form the toxic metabolites which damage pathogen DNA.
10. Macrolides including erythromycin, clarithromycin, azithromycin, spiramycin – inhibit protein synthesis in Gram-positive bacteria and some Gram-negative bacteria. In particular erythromycin is derived from *Streptomyces* spp.
11. Fusidic acid – Fusidin or fusidic acid is a product of *Fusidium* spp. and active against Gram-positive bacteria.

## 19.2 Antiviral Agents

1. Amantadine – It is normally used as a prophylactic agent against influenza virus A and given to patients with increased risk of infection, i.e., elderly or individuals suffering from other chronic infections. Amantadine interferes with the viral uncoating and viral penetration of the host cells.
2. Nucleoside analogues – These are chemical analogues of purine and pyrimidine nucleosides. Following entry into the host cell, they become phosphorylated and incorporated into growing viral genome and causes its disruption. Examples include acyclovir which is active against herpes simplex virus type 1 and type 2 and varicella zoster virus. Other derivatives include valaciclovir, famciclovir and cidofovir that are effective against these infections. Ganciclovir is highly effective against cytomegalovirus infections. Tribavirin is a synthetic nucleoside derivative active against herpes viruses, orthomyxoviruses, paromyxoviruses, Lassa virus, Rift valley virus, respiratory syncytial virus and hantavirus infections.
3. Foscarmet – It inhibits DNA polymerases of the herpes viruses and cytomegaloviruses. Although it is toxic and may produce renal failure, is effective if infection is unresponsive to acyclovir.
4. Azidovudine (AZT) – this is a potent reverse transcriptase inhibitor. This is a nucleoside and when taken up by host cells is phosphorylated to triphosphate form and resembles deoxythymidine triphosphate (dTTP), a building block of DNA. However, the host cell DNA polymerases do not use AZT but the viral reverse transcriptase enzyme incorporates AZT into growing viral DNA during reverse transcription. Since AZT lacks 3'-OH group, it cannot serve as a site for the addition of the next nucleotide and viral DNA chain is terminated.
5. Viral protease inhibitors – these include saquinavir, indinavir, ritonavir and nelfinavir.

## 19.3 Antifungal Agents

1. Polyenes – These are produced by *Streptomyces* spp. They associate with sterols (in the eukaryotic cell membranes) and rearrange sterols causing their membrane leakage resulting in cell death. Some are more toxic to fungi but others are toxic to both host and fungi membranes. These include nystatin which is used as topical treatment and particularly useful for *Candida* infections. Other compounds are amphotericin B which is highly effective against *Candida* spp. and other fungal pathogens.
2. Azoles – These are a large group of synthetic compounds which function by interfering with biosynthesis of ergosterol and thus fungal membrane dysfunction.
3. Imidazole – These include imidazole that are effective against *Candida* infections. The derivative of imidazole include clotrimazole, econazole, fenticonazole, isoconazole, miconazole, sulconazole and tioconazole. Other compounds include ketoconazole which are effective against histoplasmosis and fungal skin infections.
4. Triazoles – These include fluconazole that is effective against *Candida*, *Histoplasma* and *Cryptococcus* spp. Other compounds are itraconazole that highly effective against *Aspergillus* spp. but may also be used for *Histoplasma* and *Cryptococcus* spp.
5. Allylamines – These include terbinafine which acts by fungal cell membrane disruptions by accumulating toxic sterols in the fungal cell walls.
6. Flucytosine – This is a synthetic fluorinated pyrimidine that affects the protein synthesis of fungi and is effective against *Candida* spp. and *Cryptococcus* spp.
7. Griseofulvin, produced by fungus *Penicillium griseofulvum* is an anti-fungal agent which binds with the microtubule-associated proteins involved in tubulin assembly, thus interfering with tubulin polymerization inhibiting mitosis and other cellular functions such as secretion and transport affecting dermatophytes.

## 19.4 Antiprotozoal Agents

1. Quinine – It is effective against *Plasmodium* spp. by inhibiting haem polymerase. It also crosses blood-brain barrier and is thus effective against *P. falciparum*.
2. Chloroquine – It is effective against all four species of *Plasmodium*.
3. Mefloquine – It is effective against all four species of *Plasmodium* as well as against some strains that are resistant to other drugs. It is also useful as a prophylactic agent.

4. Primaquine – This is effective against long-term liver infections of *P. vivax* and *P. ovale.*
5. Artemisinins – This is particularly useful against serious cerebral malaria. It acts by generating free radicals that bind to parasite proteins and cause parasite damage.
6. Pyrimethamine – It is antifolate drug which is used against toxoplasmosis.
7. Atovaquone – These inhibit mitochondrial electron transport and is effective against *P. falciparum, Toxoplasma gondii* and *Pneumocystis carinii.*

CHAPTER

3

# Viruses

## 1. INFECTIOUS AGENTS: THE MISSING LINK

Infectious diseases were long thought to be the work of evil deeds of individuals or communities. Although the discovery of microbes was made by Louise Pasteur in 1860s, when he disproved the spontaneous generation theory, it was not until 1870s, when Robert Koch linked microbes as the causative agents of disease. He postulated a set of criteria which provided a foundation for the study of medical microbiology, now known as 'Koch's postulates'.

### 1.1 Koch's Postulates

Pathogens were first clearly associated with human diseases in 1870s by Robert Koch. In a series of interesting experiments, he identified *Bacillus anthracis* as the causative agent of anthrax and created a number of postulates. Koch's postulates provided a basis for advances in the infectious diseases and to prove that a specific agent is responsible for a particular disease. These postulates are:

(i) The microbe must be present in people with the disease and should be associated with the lesions of the disease.
(ii) The microbe must be isolated in pure culture from a person who has the disease.
(iii) The isolated microbe, when administered to susceptible individuals must produce the disease.
(iv) The microbe must be reisolated in pure culture from the intentionally-infected host.

However, in practice, it is difficult to fulfill all of these criteria for all human pathogens and there are exceptions. Cultures of some bacteria in the laboratory have often been a problem. For example, *Mycobacterium leprae* (known to be the causative agent of leprosy) can only be cultured in animal models. Viruses can not be cultured independently and require the host cell to reproduce.

## 2. VIRUSES

Viruses are the most common causes of acute infections and remain the biggest threat to human kind. Antibiotics may be effective in controlling bacterial diseases but are worthless against viral infections. Viruses are the smallest of the microbial world and they are observed by electron microscopy (Fig. 1). The general properties are as follows:

- Viruses have no cellular existence
- They do not grow but replicate
- Do not respire
- Do not move
- Do not fit within current 'tree of life' classification system, which is based on rRNA sequences
- Viruses are made of nucleic acid plus a protein coat

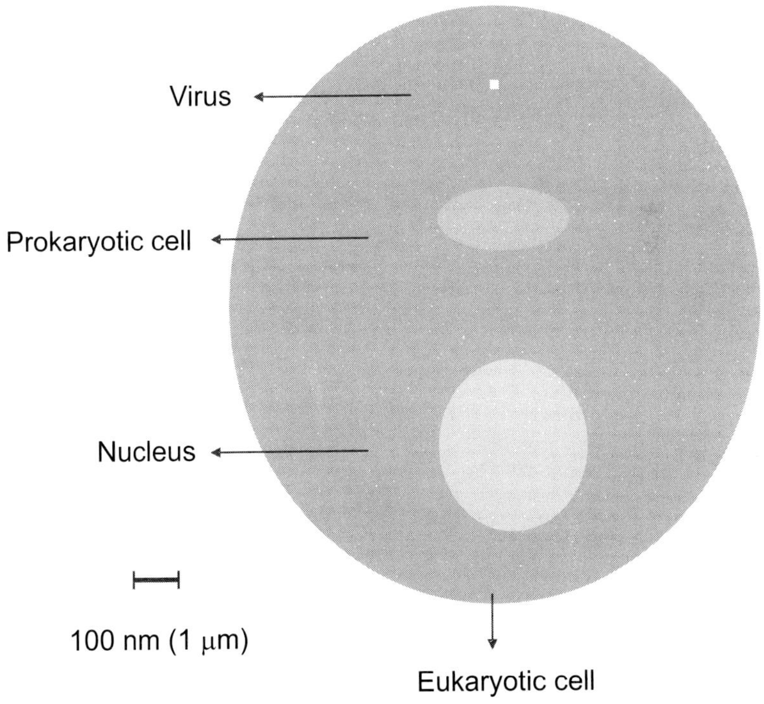


**Fig. 1** Size difference between a virus, a prokaryote and a eukaryote.

- Viruses can infect animals, plants, fungi, protists, bacteria and archea
- Most viruses are between 20 and 300 nm in size

## 2.1 Viral Discovery

The field of virology was initiated in 1796, when Jenner made use of the vaccination to eradicate smallpox. However, the first virus was discovered in the late 19th century, when Ivanowski (1892) and Beijerink (1898), worked independently with infected tobacco plants. They showed that plants were infected with an agent that could pass through filters known to remove bacteria. They proved the existence of a very small infectious agent, known as Tobacco mosaic virus. The first animal virus identified was the foot-and-mouth disease virus in 1898, while the first human virus identified was Yellow fever virus in 1901. Nearly 30,000 virus isolates have been identified which infect animals, plants, insects, bacteria, etc. These viruses are grouped into 3,600 species, 164 genera and 71 recognized families. However, there are 19 families of viruses which infect humans and are important to human health (Fig. 2).

## 2.2 Virus Structure

Viruses are nucleic acid particles, either DNA or RNA (never both), which are either single or double stranded and may be composed of a single segment (or chromosome) or multiple segments (multiple chromosomes). For example, among RNA viruses, orthomyxovirus has a single segment

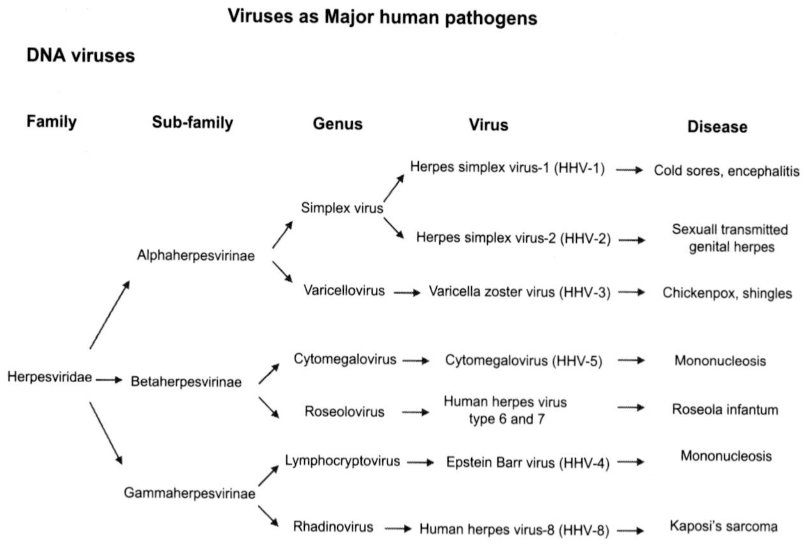

## DNA viruses

| Family | Sub-family | Genus | Virus | Disease |
|---|---|---|---|---|
| Poxviridae | | Othopoxvirus | Variola virus | Smallpox |
| | | Molluscipoxvirus | Molluscum contagiosum virus | Molluscum contagiosum |
| Adenoviridae | | Mastadenovirus | Human adenovirus | |
| Papovaviridae | | Papillomavirus | Papillomavirus | Warts, cervical carcinoma |
| Parvoviridae | | Erythrovirus | Human parvovirus (B19) | |
| Hepadnaviridae | | Orthohepadnavirus | Hepatitis B virus | Hepatitis |

## RNA viruses

| Family | Sub-family | Genus | Virus | Disease |
|---|---|---|---|---|
| Retroviridae | | Lentivirus | Human immunodeficiency virus | AIDS |
| | | HTLV retrovirus | Human T-cell leukemia virus | T-cell leukemia |
| Picornaviridae | | Enterovirus | Poliovirus | Poliomyelitis |
| | | | Coxsackievirus | Hand, foot-mouth disease |
| | | | Echovirus | Enteritis, meningitis |
| | | Rhinovirus | Rhinovirus | Common cold |
| | | Hepatovirus | Hepatitis A virus | Hepatitis |
| Togaviridae | | Rubivirus | Rubella virus | Rubella |
| | | Alphavirus | Eastern equine encephalitis virus | Encephalitis |
| Flaviviridae | | Flavivirus | Yellow fever virus | Yellow fever |
| | | | Dengue virus | Dengue fever |
| | | | West Nile virus | Encephalitis |
| | | Hepacvirus | Hepatitis C virus | Hepatitis |

**RNA viruses**

| Family | Sub-family | Genus | Virus | Disease |
|---|---|---|---|---|
| Coronaviridae | → | Coronavirus → | Human coronavirus → | Common cold |
| Calciviridae | → | Calcivirus → | Norwalk virus → | Diarrhoeal disease |
| Paramyxoviridae | → | Rubulavirus → | Mumps virus → | Mumps |
| | → | Morbillivirus → | Measles virus → | Measles |
| | → | Pneumovirus → | Respiratory syncytial virus → | Bronchitis |
| | → | Paramyxovirus → | Human parainfluenza virus-1 → | Bronchitis |
| Rhabdoviridae | → | Lyssavirus → | Rabies virus → | Rabies |
| Filoviridae | → | Filovirus → | Ebola virus → | Haemorrhagic disease |
| Arenaviridae | → | Arenavirus → | Lassa fever virus → | Lassa fever |
| Orthomyxoviridae | → | Influenza virus A → | Influenza virus A → | Influenza |
| | → | Influenza virus B → | Influenza virus B → | Influenza |
| | → | Influenza virus C → | Influenza virus C → | Influenza |
| Bunyaviridae | → | Hanta virus → | Sin Nombre virus → | Acute respiratory distress syndrome |
| Reoviridae | → | Rota virus → | Rota virus → | Diarrhoeal disease |

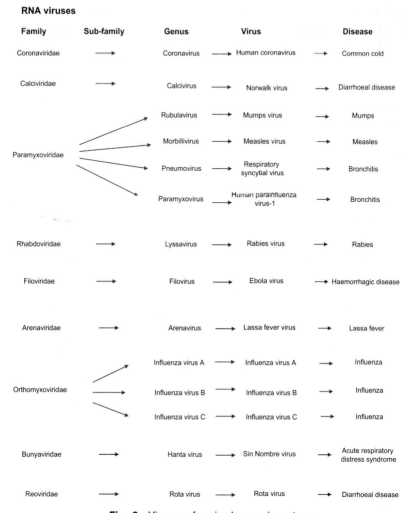

**Fig. 2** Viruses of major human importance.

(one chromosome) while the genome of reovirus contains 10 segments or chromosomes. The majority of DNA viruses have a single segment. These genomes are enclosed in a protein coat called protein capsid which protects the nucleic acid from the host nucleases and to allow virus binding to a new host cell. Nucleic acid together with the capsid is called nucleocapsid, which may be sufficient to be a complete virus and these are called 'naked viruses'. While in other viruses, the nucleocapsid is enclosed in a further lipid bilayer derived from the modified host cell membrane and these are known as 'enveloped viruses'. The envelope is also stubbed with virus-encoded proteins, which are glycosylated transmembrane proteins and appear as spikes or knobs on the surface of the virus (Fig. 3). These spikes or knobs allow viruses to infect specific cells. The morphology of the majority of viruses can be broadly described as either 'helical' in which nucleocapsids are arranged as helix or 'icosahedral' in which they are arranged into a symmetric shell (Fig. 3). However, large viruses have a complex morphology consisting both of helical and icosahedral symmetries. As viruses do not contain any cellular machinery, they are highly dependent on other cells to synthesize their components to multiply and are thus considered as acellular; also considered as 'escaped pieces of nucleic acid' and referred to as 'obligate parasites'. Thus, viruses have developed the ability to infect all living cells such as bacteria, protozoa, fungi, insects, fish, reptiles, birds and all mammals, however the scope of this book is the study of viruses which are important to human health.

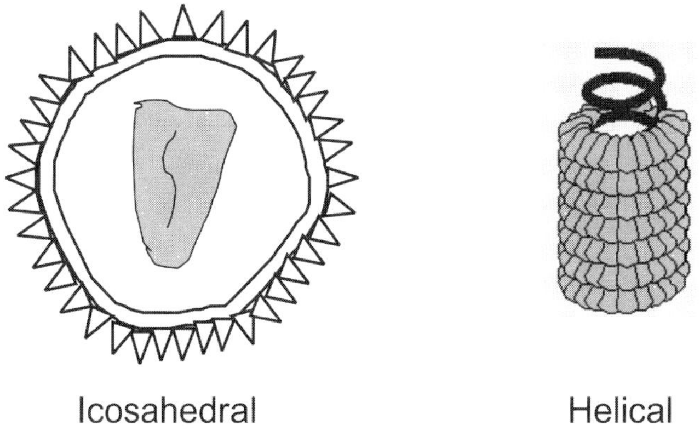


**Fig. 3** Types of symmetry in viruses.

## 2.3 Viral Propagation

Viruses are obligate intracellular parasites (they are inert outside the host cell), which grow inside the host cells using complex host metabolic machinery and biosynthetic pathways to form viral components which assemble into particles. Viruses are unable to generate energy and depend on the host cells. A complete virus particle with the ability to infect a new host cell is called a virion. The function of a virion is to infect a new host cell to continue their propagation. A virion delivers its genome (DNA or RNA) into the host cell to allow its expression (transcription and translation) by the host cell. They only contain DNA or RNA as the genetic material, which is passed to the new host cell. The DNA or RNA may be single stranded (ss) or double stranded (ds), linear or circular. If the entire genome contains only one nucleic acid segment, it is called 'monopartite genome', while genomes of viruses containing several nucleic acid segments are called 'multipartite genome'. These differences allow viruses to exhibit different replication strategies. Of interest, the genomes of viruses vary in size from 3,500 nucleotides in small phage (virus) to 280 Kbp in some herpes viruses.

## 2.4 Viral Infection of the Host Cell

A complete infectious cycle includes attachment of the virion to the host cell followed by their entry/uptake, production of viral mRNA and proteins, genome replication and assembly and release of new particles (Fig. 4). These can be divided into the following steps:

### *2.4.1 Tropism*

Viruses do not infect all cells and their ability to infect is usually limited to certain cells, tissues or certain species and this property is referred to as cell-tropism, tissue-tropism or species-tropism. This is due to the presence of specific receptors or specific physiological conditions or the status of the

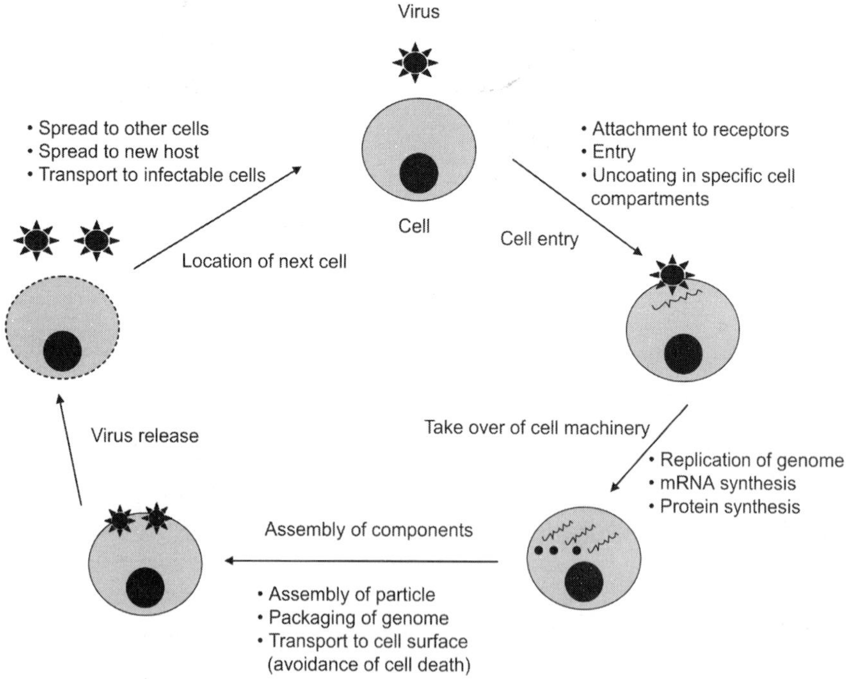


**Fig. 4** Viral infection of the host cell.

host's defence. For example, rhinoviruses do not grow at temperatures exceeding 34°C and thus only infect cells in the outer layer of the nasal mucosa and are unable to invade into the deeper tissues where the temperatures are high. The rabies virus specifically binds to acetylcholine receptor expressed on neurones and HIV specifically binds to CD4 receptors expressed on T-lymphocytes as specific receptors. However, there are exceptions such as the St. Louis encephalitis virus which can infect different species.

### 2.4.2 *Attachment*

As indicated above, viruses usually do not infect all cells and must come in contact with cells or tissues that they have evolved to infect. The protein coat of the virus protects it from the cellular enzymes such as DNAses and RNAses and at the same time it allows the attachment of the virion to the membrane of the host cell. Binding of the virion to the host cell is highly specific and they only can bind to certain cells. This specificity is provided due to the presence of specific protein(s)/glycoprotein(s) in the coat of the virion and the presence of specific receptors on the host cells. The cellular receptors are usually glycoproteins, sialic acid or heparin sulfate. For example, HIV binds specifically to T-lymphocytes expressing CD4 receptors.

### 2.4.3 *Entry*

Following attachment, the virus traverses the lipid bilayer surrounding the host cell, without killing the cell. This is an energy-dependent process, which can be divided into the following categories:

### 2.4.4 *Receptor mediated endocytosis*

Non-enveloped viruses lack the lipid membranes found in the enveloped viruses and cannot enter via a simple process of membrane fusion between virus envelope and the host cell membrane. These viruses bind specifically to host cell receptor(s) and are internalized in an endocytic vacuole. The virus then disassembles in acidic environment of early endosome (~ pH 6) triggering endosome lysis, thus mediating an escape of partially disassembled core particles into the cytosol. However, some viruses such as reovirus do not escape from the endosome. They rely on enzymes found in lysosomes (i.e., proteases) to carry their uncoating. Lysosomal proteases convert reovirus particles into an infectious sub-viral particle (ISVP) which then penetrates into the cytoplasm.

### 2.4.5 *Translocation*

Virion entry into the host cell may include translocation across the host cell plasma membrane. This mechanism is normally used by non-enveloped viruses.

### 2.4.6 *Surface fusions*

Some viruses fuse directly with the host cell membrane, resulting in uncoating of the viral genome at the cell membrane (Fig. 5). The envelope

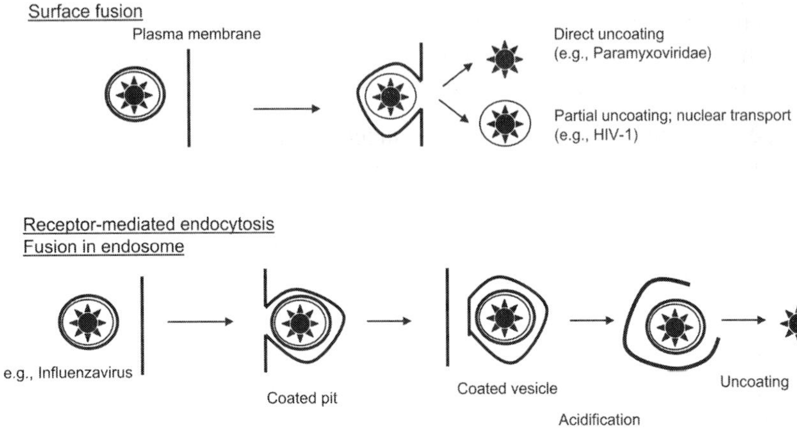


**Fig. 5** Surface fusion of viruses with the host cells.

remains in the plasma membrane and the internal components are released into the cytoplasm. This mechanism is normally used by enveloped viruses and requires the interaction of specific viral proteins in the envelope with proteins in the host cell membrane.

### *2.4.7 Pore formation*

Some viruses form pores in the host cell membrane. For example, poliovirus, after binding to its receptor forms a pore in the host cell membrane, through which the viral RNA is then released into the cytoplasm.

### *2.4.8 Uncoating*

Once inside the cell, the virus disassembles itself (e.g., triggered by pH changes, enzymes, etc.), alone or with the aid of cellular enzymes. This disintegration takes place in a way that their genetic information and any associated enzymes remain intact and viral nucleic acid and associated enzymes are directed to the appropriate cellular compartment. It is important to note that the ultimate purpose of a virus is to produce viral mRNA that can be used to translate viral proteins. Viruses which do not enter nucleus, do not have access to host transcriptases in the nucleus and thus they must possess their own viral transcriptases to generate mRNA.

### *2.4.9 Transport to the nucleus*

For DNA viruses, the genome (viral DNA) is released in the cytoplasm and then it is transported to the nuclear pore. This process is dependent on the presence of nuclear localization signals (NLS, short sequences) within the proteins associated with the viral genome traversing through the nuclear pores. Interactions between NLS and cytoplasmic NLS receptor(s) allow their docking to the nuclear pore. Translocation across the pore requires the action of additional proteins, including a small GTP-binding protein known as Ran, resulting finally in the release of viral DNA into the nucleus, e.g., Adenovirus, herpes virus and papillomavirus.

### *2.4.10 Replication and viral protein production*

These normally require viral proteins and some host proteins, which are broadly divided into two categories:

i. Early proteins, which control the next phase of replication cycle e.g., genome replication.
ii. Late proteins, which are usually structural proteins, e.g., capsid.

**2.4.10i Bidirectional replication from a circular substrate** Replicate using a specific origin of DNA replication which serves as an initiation point for bidirectional DNA replication.

**2.4.10ii Replication from a linear substrate** First one strand is made in its entirety and then the next strand is made e.g., Adenovirus.

**2.4.10iii Replication via an RNA intermediate** For example, hepatitis B virus - a dsDNA genome is converted into an RNA by virion enzyme reverse transcriptase during the virus life cycle.

### *2.4.11 Assembly and release*

Once viral components have been synthesized, they are packaged into virions by viral proteins. Finally different components are assembled together and virus particles are released from cells by lysis of cells or budding from the cell membranes, usually the cytoplasmic membrane. Both processes lead to changes in the cell membranes, which can often be observed *in vitro*, as part of the virally-induced cytopathic effects. However, these processes differ between enveloped and naked viruses. For example, naked viruses are assembled and matured intracellularly to become fully infectious virions. This is also true for viruses that damage their host cells during their release process and thus they must complete their maturation and achieve infectivity while intracellular. In contrast, in the enveloped viruses, assembly and release occurs simultaneously. To achieve this, viral proteins are inserted into the host cell plasma membrane and nucleocapsid bind to these proteins on their cytoplasmic domains and wrapped up in the membrane (Fig. 6). This leads to virion extrusion or budding out into the external environment. In some viruses, this is followed by modification of viral surface proteins (maturation by proteolytic cleavage and rearrangement). This process is not dependent on viruses to lyse the host cells to be released. However, viruses using this mechanism of assembly and release may have cytolytic effects on the host cells (e.g., paramyxoviruses, rhabdoviruses) or may simply use host cell machinery

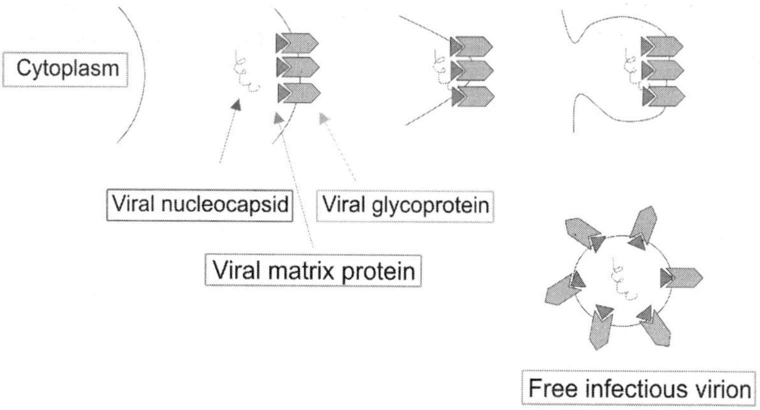


**Fig. 6** Assembly, maturation and release of enveloped viruses. Virus particles are released from cells by lysis of cells of budding from cell membranes, usually the cytoplasmic membrane. Both processes lead to changes in cell membranes, which can often be observed in vitro, as part of the virally-induced cytopathic effect.

for the replication without causing any damage (e.g., retroviruses). For example, in the case of HIV infection, CD4 T-cells are not killed by the virus but are targetted by the immune cells due to the expression of viral proteins (gp120) on their surface. However, there are exceptions. For example, the herpes virus is an enveloped viruses but their assembly and maturation occurs in the nucleus and in the nuclear membranes and their effects are cytolytic on the host cells.

## 3. VIRAL CLASSIFICATION

The origin of viruses is unclear as they co-evolve with their hosts. We are only aware of the existing viruses or the ones that evolved a few decades ago. Generally, viruses are classified based on their genome structure and replication strategies into following broad categories.

### 3.1 RNA Viruses

These comprise more than 70% of the known viruses. Viral RNA genome can act as its own message (positive strand viruses) or the complementary strand can be the mRNA (negative strand viruses). All RNA viruses except retroviruses encode RNA-dependent RNA polymerase. In the negative strand RNA viruses, this polymerase must enter the cytoplasm along with the viral genome. This is necessary in order for the virus to generate mRNA from its genome. All RNA viruses replicate in the cytoplasm except orthomyxoviruses (Influenza), hepatitis delta virus, borna disease virus and retroviruses. And because they replicate in the cytoplasm and do not have access to the proof reading machinery in the nucleus, they have a high mutation rate than the DNA viruses. This leads to the emergence of diverse virus variants and may explain why a majority of viruses are RNA viruses.

### *3.1.1 Single stranded RNA viruses*

**3.1.1i Single stranded RNA viruses (ssRNA) – positive sense (plus sense)** Following entry into the host cytoplasm, plus sense RNA binds to ribosomes and directly translates viral proteins. However, this is only a single molecule and to expedite this process, plus sense RNA is used as a template to synthesize complementary minus sense RNA by viral polymerase. The minus sense RNA is then used as a template to make more plus sense RNA. The increased numbers of plus sense RNA are then used as mRNA to make more minus sense RNA and more viral proteins.

1. Functions as mRNA and can directly initiate translation for viral proteins.
2. Translation of virion RNA as mRNA.
3. Synthesis of (-)sense RNA on (+)sense template.
4. Synthesis of (+)sense RNA, mRNA and (-)sense RNA.
5. Translation of (+)sense and mRNA, synthesis of structural protein.
6. Assembly of structural proteins and (+)sense RNA and maturation of virions e.g., *Leviviridae, Togaviridae.*

**3.1.1ii Single stranded RNA viruses (ssRNA) – negative sense (minus sense)** These viruses can not function as mRNA and are hence referred to as antisense or minus sense. Their genomes must first be transcribed to mRNA but they do not enter the nucleus. Thus, all minus sense RNA viruses contain a transcriptase in addition to their genome. Following entry into the host cell, the released minus sense RNA molecule serves as a template and transcribes into a plus sense RNA by viral transcriptase, which is used as a mRNA to synthesize the viral genome and its proteins. Overall, this process is totally dependent on viral transcriptase and the viral RNA alone is not infectious to the host cell.

1. Cannot function as mRNA but acts as complementary to the positive sense strand for viral protein translation.
2. Primary transcription of virion (-)sense RNA in cytoplasm, producing mRNA and (+)sense RNA.
3. Translation mRNAs, accumulation of products.
4. Production of full-length (+)sense RNA and thus of genomic (-)sense RNA.
5. Secondary transcription from progeny (-)sense RNA, translation, accumulation of structural proteins.

6. Nucleocapsid assembly and maturation, budding of nucleocapsid through the host membrane containing viral envelope proteins e.g., *Paramyxoviridae, Filoviridae.*

**3.1.1iii Retroviruses [single stranded RNA viruses replicating via dsDNA] (diploid ssRNA viruses)** The genome of these viruses is unique and is composed of two identical, plus sense ssRNA with each 7 – 11 kb in size that are non-covalently linked. Technically, they have a single segment (one chromosome), as both strands are linked. In addition, virion contains RNA-dependent DNA polymerase (reverse transcriptase), transfer RNA that serves as a primer and ribonuclease that degrades the viral genome. Upon infection of the host cell, plus sense ssRNA serves as a template and binds to the complex, i.e., viral reverse transcriptase and tRNA. This leads to the synthesis of a complementary DNA molecule. Next the DNA molecule is used as a template to make a complementary DNA strand using viral polymerase to form a linear dsDNA. This DNA is then translocated into the nucleus and integrated into the host genome at random with the aid of viral integrase sites. After a period, the viral DNA can then be transcribed by the host RNA polymerase to produce new viral RNA genome and viral proteins. After assembly and budding, structural proteins of the immature virions are processed proteolytically and form a mature, complete infectious virion.

1. Reverse transcription in the cytoplasm by virion-associated reverse transcriptase into intermediate RNA/DNA complex.
2. Conversion of RNA/DNA complex into the linear and circular proviral dsDNA; import into the nucleus.
3. Integration of linear proviral DNA into host cell DNA, by reverse transcriptase.
4. Replication and transcription using host enzymes.
5. Modification of transcription leading to (+)sense RNA.
6. Translation, accumulation of structural protein into viral nucleoprotein, budding through the membrane containing viral envelope glycoproteins e.g., *Retroviridae* including HIV-1.

### *3.1.2 Double stranded RNA viruses*

The genome of these viruses normally consists of several segments (multiple chromosomes). Upon infection of the host cell, the viral genome is transcribed by the viral polymerase (usually this step is carried out within the partially open virion). This results in the formation of plus sense RNA (or mRNA), which are released into the host cell cytoplasm and translate into viral proteins. Next the assembly of plus sense RNA and viral proteins takes place resulting in the formation of immature virions. While in the immature virions, plus sense RNA serves as a template for the synthesis of the complementary strand resulting in double stranded RNA segments.

1. Primary transcription in the virion core in the cytoplasm and export of (+)sense RNA to the cytoplasm.
2. Translation of (+)sense RNA, accumulation of viral proteins.
3. Assembly of (+)sense RNA and viral proteins into immature virions
4. Transcription of (+)sense RNA into dsRNA in virions.
5. Secondary transcription of dsRNA.
6. Final assembly/maturation of virions e.g., *Hypoviridae, Reoviridae.*

## 3.2 DNA Viruses

The majority of DNA viruses contain linear dsDNA but some have circular DNA. DNA viruses are normally transcribed and replicated in the nucleus and thus utilize host enzymes to generate mRNA. Since transcribed and replicated in the nucleus, they also have access to host proof reading machinery and exhibit lower mutation rates as compared to the RNA viruses. The genome alone of these viruses is sufficient to infect the host cell. However, there are exceptions, where DNA virus transcription takes place in the cytoplasm using viral enzymes similar to the RNA viruses.

### *3.2.1 Single stranded DNA viruses*

Replication requires formation of a 'replicative form' (RF) double-stranded DNA intermediate, which is formed soon after infection, and carried out by the host cell DNA polymerases.

1. Conversion into dsDNA.
2. Early transcription (by host enzymes).
3. Translation of (regulatory) protein and 'rolling circle' ssDNA replication.
4. Late transcription (usually mediated by viral proteins).
5. Translation of late (=structural) proteins.
6. 'Sequestering' of viral genomic ssDNA.
7. Assembly into virions e.g., *Paroviridae, Microviridae.*

### *3.2.2 Double stranded DNA viruses*

The genome of dsDNA viruses serves as a template for mRNA and for self-transcription.

1. Replication process.
2. Primary transcription by the host enzymes.
3. Translation of early (=regulatory) proteins.
4. Viral genomic DNA replication (usually by the host enzymes).
5. Late transcription (usually mediated by the viral proteins).
6. Translation of late (=structural) proteins.
7. Assembly of structural proteins and DNA into virions, e.g., *Herpesviridae, Adenoviridae.*

### 3.2.3 *dsDNA replicating via ssRNA*

1. Entry of viral DNA into the nucleus.
2. Transcription by host RNA polymerases into mRNA and (+)sense RNA.
3. Translation of mRNA and (+)sense RNA in the cytoplasm.
4. Interaction of viral proteins with (+)sense RNA, assembly of provirions and conversion of RNA to RNA/DNA complex.
5. Conversion of RNA/DNA complex to dsDNA.
6. Virion maturation and budding e.g., *Hepadnaviridae*.

## 4. VIRAL GENETICS

Viruses are simple particles that are dependent on other cells for their existence. So simple, that they are referred to as "escaped pieces of nucleic acids". Their genome may consist of just few genes in small viruses to as many as 200 genes (up to 200 kbp) in large viruses. Due to their simplicity, they do not have the ability to ensure that their genome is conserved in the replication cycle. Thus, their replication cycles result in mutations and the emergence of new variants. In addition, genomes from different viruses can be recombined to create new viruses. Both processes may result in the modifications of the functional properties of viruses, which may make them more virulent to their hosts or more adaptable to different environmental conditions.

### 4.1 Mutations

Mutations in viral genomes are the single most challenging aspects in the viral pathogenesis. The ability of viruses to change, allow them to evade the host immune system making the existing antibodies obsolete. Since DNA viruses enter host cell nucleus, they have access to the proof reading machinery. Thus, the rate of mutation in DNA viruses is similar to what is observed in eukaryotic cells ($10^{-8}$ to $10^{-11}$ errors per incorporated nucleotide) and thus relatively low. In contrast, RNA viruses do not enter host cell nucleus and therefore do not have the access to host cell machinery and thus their rate of mutation is much higher ($10^{-3}$ to $10^{-4}$ errors per incorporated nucleotide), therefore they generate mutants (new variants) more frequently. Viral mutations occur frequently but they are not necessarily kept within the viral genomes. The majority of mutations that affect viral infectivity or their replication are lost. It is the mutations that do not affect essential viral functions which are preserved in viral genomes.

## 4.2 Recombinations

This term is used when genomes of two different viruses are joined. This normally occurs when two different viruses belonging to the same viral type (e.g., two influenza viruses) infect the same host cell and interact during replication and produce daughter cells which have some genes from both parent viruses. Recombination is divided into two types.

## 4.3 Recombination by Independent Assortment

This occurs in viruses with multiple segments (multiple chromosomes). Once the genomes are available in the same host cell, various fragments can be recombined. The result is a new variant with major antigenic changes, a process known as 'antigenic shift'. The new variants may be able to infect animals which are immune to the parent viruses. Antigenic shifts in influenza virus are now well-established to produce devastating effects on human health resulting in pandemics. For example, the influenza virus strain H1N1 was responsible for the 1918 to 1919 influenza pandemic, which caused 20 million deaths and then disappeared. This strain reappeared (due to recombinations) in 1934, then 1947, and again in 1977.

## 4.4 Recombination of Incompletely Linked Genes

This is described by the recombination of genes present on a single segment (single chromosome). This normally occurs in genes that segregate together and are hence called 'linked genes'. However, due to the recombination, the linkage is broken and is now referred to as 'incompletely linked genes'. This is usually observed in DNA viruses in a process similar to prokaryotes and eukaryotes. DNA strands are broken and rejoined to DNA strand of a different molecule resulting in a recombinant virus. These processes result in the creation of novel viruses with modified virulence and infectivity.

## 4.5 What Causes Mutations

Mutations are produced by different factors:

1. Spontaneous, such as polymerase error as described above.
2. Physically induced
    - UV light, a problem if no access to repair
    - X-rays
3. Chemically induced

### 4.6 Mutations to our Benefit – Vaccines

Mutations occur in viruses to make them more adaptable and robust. This property has been used for a long time for the benefit of human kind against several diseases in the form of vaccination. For example, some viruses that produce human infections can be used to infect other mammalian cells, e.g., monkey cells. The persistent culture of these viruses with monkey cells will produce mutations, which may allow them to become more adaptable to infect monkey cells than human cells. These variants will be more infectious to monkey cells but at the same time will have several features (proteins/glycoproteins) consistent with the parent virus (i.e., human virus) and may be used as a vaccine for humans. For example, the vaccine strains of the polio virus were developed by growing polio viruses in monkey kidney cells. After persistent culturing, mutations and selections produced variant polio viruses that were highly infectious to monkey kidney cells. Since mutations also affected genes coding for viral coat proteins. These variant viruses were able to bind to the human intestinal cells but were unable to bind to the human neural cells. Thus, these variant became a valuable tool for vaccination as they were unable to produce disease (paralysis) but bound to intestinal cells, which resulted in immunity against these viruses. Some variants are stable as they possess several mutations in the gene coding for coat protein but others are unstable and may revert to the original virus. These parameters need to be measured carefully to ensure appropriate vaccines with minimal side effects. Similarly, recombinations can be used to develop new vaccines. In addition, viruses can be used to carry specific genes to target specific cells for gene therapy.

## 5. VIRAL INFECTIONS

There are six basic types of viral infections which are as follows:

### 5.1 Acute Lytic Infections

- Viral infection produce infectious progeny in the host cell resulting in cell death.
- Clear clinical signs of disease but the duration of disease is short
- Transmission of the virus must be efficient, e.g., mumps, measles.

### 5.2 Sub-clinical Infections

- Like acute infections, sub-clinical infections produce infectious progeny and some cell death, but no symptoms.
- All other characteristics are shared with acute infections.
- 99% of polio infections are sub-clinical.

## 5.3 Persistent Infections

In these infections, the virus is not cleared from the human body and is maintained in the host cells for longer periods of time. These are several types of persistent infections that differ in their mechanisms and are described below.

### *5.3.1 Chronic Infection*

In these infections, the virus targets their host cells and replicates inducing a low level of symptoms. However, new virions are controlled by innate and acquire host immune system (in particular, antibodies and T-cells). In chronic infections, there is a continued presence of infectious virions. This is associated with long incubation periods followed by progressive disease with the clear onset of clinical symptoms, e.g., Hepatitis B virus.

### *5.3.2 Latent Infections*

In latent infections, the virus produces an acute infection and becomes latent with limited replication. The virus is maintained in non-dividing or rarely dividing cells, e.g., Herpes simplex virus. Viruses are usually not detectable during the latent period, however, they are able to reactivate themselves in response to diverse signals (e.g., heat, ultraviolet irradiation, immune suppression, chemotherapy, etc.).

### *5.3.3 Tumourigenic Infections*

In these infections, viruses may or may not be released from the host cells. Viral proteins inactivate host cell tumour suppresser proteins (e.g., Rb, p53) resulting in impaired cell cycle. Examples include T antigens of papova viruses, E6, E7 of papilloma viruses, and Epstein-Barr virus. This results in the transformation of the cell to become immortalized. These, transformed cells, have reduced requirement for growth factors and become cancerous, i.e., grow forever.

### *5.3.4 Progressive Infections*

These infections take years to show any clinical symptoms. They normally begin with influenza-like symptoms and become latent. During the latent period, they propagate infectious particles. Once reactivated, they overwhelm the immune system, e.g., HIV.

# 6. VIRAL PATHOGENESIS

The process of viral infection leading to disease is called viral pathogenesis. Our ability to understand the basic pathogenic mechanisms associated with viral diseases both at the cellular and molecular level are crucial for identifying therapeutic targets. As indicated above, there are hundreds of human viruses causing diverse diseases. The description of each virus with their mode of action is beyond the scope of this book. Here, HIV has been used as a model virus to describe various features associated with its clinical and non-clinical aspects.

## 7. CONTROL OF VIRAL INFECTIONS

Infections due to viruses vary from acute fatal infections that last for a few days to chronic infections that last for years with varied severity. There is no single parameter that could be used to control all viral infections and each virus requires a different set of measures to help control its infection. Below, the most widely, known measures are indicated for the control of viral infections. It is important to note that since viruses depend on the host cells for their survival, it is unlikely that host killing is the objective, and host death is most likely a result of the side effects. Thus, viruses usually produce mild infections with the obvious exceptions of HIV, ebola virus, rabies virus, hanta virus, etc. However, the extremely close association of the virus and its host means that any consideration of treatment of therapeutic interventions must be viewed with this in mind. The ability of viruses to use cell machinery and biosynthetic processes of the hosts to multiply, make them very difficult to target. The identifications of specific targets in virus-infected host cells is the key for the development of specific therapeutic measures, i.e., antivirals. A clear understanding of the viral binding to the host cells, spread within the host, exit from the body and associated mechanisms is crucial for the correct diagnosis and to develop therapeutic interventions and to identify targets for vaccines development.

### 7.1 Immunoprophylaxis

The use of host immune system to control infections is called immunoprophylaxis. These include the use of vaccines to stimulate host immune response (active immunity) or the use of antibodies to provide protection against a specific viral disease (passive immunity). The types of viral vaccines are described below:

#### *7.1.1 Attenuated live viruses*

These include viruses which have been attenuated by laboratory manipulations. These viruses have the ability to infect the host but are unable to produce disease or have minimal side effects but evoke an effective immune response that can provide protection against future exposure to the virus. The immunity provided using this type of vaccination is life long.

#### *7.1.2 Killed (inactivated) viruses*

These vaccines include whole or some component of viruses killed by chemical or physical means. These viruses can not produce infection and have minimal side effects, however they do not produce lifelong immunity.

### *7.1.3 Recombinant produced antigens*

In this vaccination, an important antigen is identified and its gene is cloned and the protein is produced in an artificial system. This protein (antigen) is called recombinant antigen and can be used as a vaccine.

The recombinant antigen is given to the susceptible populations and it induces humoral response (antibody production) as well as a cell-mediated response, i.e., activation of T-lymphoctes. This is referred to as 'active immunization'. The major problem in active immunization is that it only works before the susceptible hosts are exposed to the virus. In contrast, passive immunization is achieved when antibodies (formed in another host) are injected into the infected patient. Passive immunization is used for patients already infected with the virus and where vaccination is obsolete.

## 7.2 Chemotherapy in Viral Infections

The major problem in the development of antiviral compounds has been the identification of biochemical or molecular differences between the normal cell and the virus-infected cell. Previously, it was thought that there were no differences, which halted any research in the identification of chemotherapeutic approaches for viral infections. However, now we know that some steps in the replication of some viruses differ considerably between the infected and the non-infected cells or some viruses use their own enzymes (polymerases, integrases, nucleases, etc.) to carry out their replication. These enzymes may differ both structurally and functionally between the viral and host enzymes. The identification of these and other steps in the replication of viruses in the host cells offer potential in the development of chemotherapeutic approaches. Although the future looks promising, to date, there has been limited if any success in the rational drug design based on the viral antigen structure or other targets and the majority of compounds have been identified based on their antiviral activity. The antiviral compounds can directly act on viral targets (such as amantadine, rimantadine, foscarnet and viral protease inhibitors) or first become activated intracellularly by phosphorylation before targetting viral antigens (such as acyclovir, ganciclovir, penciclovir, bromovinyldeoxyuridine, zidovudine and dideoxynucleoside analogs). However, the majority of these compounds have toxic side effects on the host cells and more selective compounds are needed for targetted and effective therapy. Other problems are that the majority of these compounds are active only against the replicating virus and useless against the latent phase of the infection. For example, there are no compounds that can target the latent phase of HIV infection and they are only effective once HIV starts replicating. Also, there is clear evidence that viruses develop drug resistance, thus the existing or future anti-viral compounds should be used with caution. Based on these, three approaches can be used to target viral infections as described below:

1. Antiviral agents that directly inactivate viruses.
2. Antiviral agents that inhibit viral replication.
3. Immunomodulators, which boost the host immune response.

Other approaches include avoidance of viral exposure. For example, HIV prevention can be achieved by avoiding sexual contact with the infected individuals or exposure to the contaminated blood or the use contaminated needles.

Other control measures include reduction of viral exposure to the susceptible hosts by eliminating virus reservoir, eliminating the vector and by improving sanitation. For example, the yellow fever virus is transmitted by the bite of mosquitoes. The control of mosquitoes have reduced the risk of yellow fever transmission. This is achieved by the introduction of procedures that reduce mosquito populations or limits the access of the vectors to humans by draining swamps or standing water, use of insecticides, screening homes, and using insect repellants or protective clothing or insecticide-impregnated nets.

Other control approaches include improved sanitation. This method is especially useful for the control of viruses that are transmitted by the faecal-oral route. For example, hepatitis A virus is transmitted by poor sanitary conditions.

It is the combinations of these approaches that should be useful in the control of viral infections.

## 8. MAJOR VIRAL PATHOGENS OF HUMANS

### 8.1 DNA Viruses

#### *8.1.1 Organism: Herpes virus*

There are different herpes viruses that can cause human infections including herpes simplex virus (HSV), varicella zoster virus (VZV), cytomegalovirus (CMV) and Epstein Barr virus (EBV) as described below. Their general properties include:

#### *8.1.2 Organism: Human herpes simplex virus (HHSV)*

Biology: HSV is an enveloped, double stranded DNA virus with a genome size of approx. 150 kbp and is around 100 nm in size. There are two types of viruses, HSV-1 and HSV-2.

**Disease:** HSV causes sores or blisters on the lips, mouth (usually HSV-1) and the genital areas (usually HSV-2) with fever. Other symptoms may include malaise, pain and muscleache. It can also cause eye infections known as herpes simplex keratitis.

**Diagnosis:** Diagnosis is made on physical examination.

**Treatment:** Treatment includes the use of acyclovir (Zovirax), famciclovir and valacyclovir.

**Transmission:** Sexual transmission or contact with the contaminated secretions.

**Occurrence:** Worldwide.

### *8.1.3 Organism: Varicella-zoster virus (VZV) also known as human herpes virus-3 (HHV-3)*

**Biology:** VZV is an enveloped, double stranded DNA virus with a genome size of approx. 125 kbp and is around 100 nm in size.

**Disease:** It is the causative agent of chickenpox and shingles. Chickenpox is a skin infection (during childhood) with the appearance of a large numbers of itchy, fluid-filled blisters. The disease may be associated with fever, headache and loss of appetite. While shingles is a painful rash with a burning sensation and may be associated with fever, malaise, headache and red skin which may be secondary to chickenpox.

**Diagnosis:** Based on physical examination and clinical symptoms. For atypical symptoms, serology-based assays are employed for sensitive, specific and rapid diagnosis.

**Treatment:** Treatment may not be necessary for skin infection, but is important if the eyes, face, mouth or genitals are involved. The antiviral agent, acyclovir, famciclovir or penciclovir can be used to treat these infections.

**Transmission:** Contact with the contaminated secretions or sexual transmission.

**Occurrence:** Worldwide.

### *8.1.4 Organism: Cytomegalovirus (CMV) also known as human herpes virus-5 (HHV-5)*

**Biology:** CMV is an enveloped, double stranded DNA virus with a genome size of approx. 230 kbp and is around 100 nm in size.

**Disease:** It may produces infection with symptoms such as mononucleosis-like syndrome, fever and possibly mild hepatitis (especially during pregnancy).

After a brief initial infection, the virus remains dormant for life. However, severe impairment of the body's immune system may reactivate the virus from the dormant state.

**Diagnosis:** Usually it is not diagnosed as an infection and may be asymptomatic, but ELISA methods are available for demonstration of antibodies against the virus.

**Treatment:** There is no specific treatment, but anti-viral ganciclovir can be used for infants and immunocompromised patients.

**Transmission:** Oral, respiratory route as well as sexual transmission.

**Occurrence:** Worldwide.

### 8.1.5 *Organism: Roseolovirus also known as human herpes virus type 6 and 7 (HHV-6 and HHV-7)*

**Biology:** It is an enveloped, dsDNA with genome size of 200 kbp and is around 120–200 nm in size.

**Disease:** It causes exanthem subitum (also called roseola infantum), in infants and is associated with high-grade fever for few days and the appearance of a rash with pink papules (roseola meaning pink-coloured rash) and is mainly distributed on the trunk, arms and neck and fades within a few days with the development of antibodies. As with other herpes viruses, it remains dormant due to the effective immune response but can be reactivated upon impairment of the immune response.

**Diagnosis:** Diagnosis is usually not required but can be made by the demonstration of viral antibodies in the blood.

**Treatment:** The disease is self limiting but acetaminophen and ibuprofen can be given to relieve fever and pain.

**Transmission:** The virus is found in the saliva of a majority of adults and may serve as a primary source of transmission.

**Occurrence:** Worldwide.

### 8.1.6 *Organism: Epstein-Barr virus (EBV) also known as human herpes virus-4 (HHV-4)*

**Biology:** EBV is an enveloped, double stranded DNA virus with a genome size of approx. 172 kbp and is around 100 nm in size.

**Disease:** The causative agent of infectious mononucleosis which may be associated with fever, lymphadenopathy, a sore throat, malaise, tiredness, splenomegaly, hepatomegaly and abnormal liver function.

**Diagnosis:** Diagnosis is made using serology-based tests.

**Treatment:** The disease is self limiting but antiviral agent, acyclovir can be used to treat this infection.

**Transmission:** Contact with the contaminated secretions or sexual transmission.

**Occurrence:** Worldwide.

### *8.1.7 Organism: Rhadinovirus, also known as human herpes virus-8 (HHV-8)*

**Biology:** EBV is an enveloped, double stranded DNA virus with a genome size of approx. 160 kbp and is around 150-200 nm in size.

**Disease:** It is associated with Kaposi's sarcoma. These tumours are usually formed in people with AIDS and associated with purplish skin and can also cause lymphomas (lymph nodes tumours).

**Diagnosis:** Diagnosis is made using serology-based tests.

**Treatment:** The use of antiviral agents such as acyclovir, foscarnet, ganciclovir and cidofovir are effective.

**Transmission:** Contact with the contaminated secretions or sexual transmission.

**Occurrence:** Worldwide.

### *8.1.8 Organism: Variola virus (smallpox)*

**Biology:** It belongs to the poxviridae family (including cowpox virus, camel pox, monkey pox, molluscum contagiosum virus, etc.). They are the largest known DNA viruses and replicate entirely in the cytoplasm of infected cells and thus do not require nuclear factors for replication.

The virion contains DNA-dependent RNA polymerase and several other enzymes required for their replication. It is an enveloped, double-stranded DNA virus with approx. 186 kbp genome and is around 200 – 300 nm in size.

**Disease:** The causative agent of smallpox, which is characterized by the appearance of lesions on the oral mucosa or palate and spread to the face, forearms, hands and finally to the rest of body. The disease is associated with high fever, malaise, headache, vomiting with a mortality rate of more than 94% in unvaccinated individuals.

**Diagnosis:** The disease can be diagnosed by the demonstration of viral antibodies or viral antigens in the blood using ELISA.

**Treatment:** Vaccination is the best option. The antiviral agent, cidofovir may be effective.

**Transmission:** Smallpox is highly contagious and infected patients must be quarantined.

**Occurrence:** The world was declared free of smallpox by the World Health Organization in 1980.

### *8.1.9 Organism: Molluscum Contagiosum virus (MCV)*

**Biology:** MCV is an enveloped, double stranded DNA virus with approx. 185 kbp genome and is around 200 – 320 nm in size.

**Disease:** It causes molluscum contagiosum infection, which is small, pink-coloured spots which appear on the face, arms and legs of infected individuals, especially in children. If the infection is sexually transmitted, the spots can be seen on the genitals and thighs.

**Diagnosis:** Based on clinical symptoms as well as serology-based assays.

**Treatment:** There is no recommended treatment as the disease disappears within a few weeks but may require cryotherapy or removal of spots or topical application of cidofovir in severely infected individuals.

**Transmission:** The infection is normally transmitted through direct contact with infected persons or indirectly by sharing towels or from swimming pools.

**Occurrence:** Worldwide.

### *8.1.10 Organism: Human adenovirus*

**Biology:** It is naked, double-stranded DNA virus with the genome size of 350 kbp and is around 90 nm in size.

**Disease:** They are normally associated with respiratory illness but can cause gastroenteritis, conjunctivitis, cystitis, and rash. Respiratory infections are associated with symptoms of the common cold syndrome, pneumonia, croup, and bronchitis.

**Diagnosis:** Based on clinical symptoms and by demonstrating viral antibodies and viral antigens in the blood.

**Treatment:** There is no recommended treatment (the disease is usually mild). Drugs are targetted for symptoms.

**Transmission:** The infection is normally transmitted through direct contact with infected persons or indirectly by sharing towels or from swimming pools.

**Occurrence:** Worldwide.

### *8.1.11 Organism: Human Papillomavirus (HPV)*

**Biology:** These are double-stranded viruses with DNA genome (approx. 8 kbp). They are non-enveloped and are around 52 – 55 nm diameter in size.

**Disease:** It is a common cause of sexually transmitted infection. They are usually harmless, such as genital warts, lumps in the genital areas including the vagina, cervix, vulva (area outside of the vagina), penis and rectum without any symptoms. But these infections can lead to cancers of the cervix, vulva, vagina, anus or penis.

**Diagnosis:** The diagnosis is made on physical examination. Applying vinegar (acetic acid) to the affected areas makes it white and thus more visible.

**Treatment:** Cryotherapy.

**Transmission:** Sexually transmitted infection.

**Occurrence:** Worldwide.

### *8.1.12 Organism: Human parvovirus B19 (HPV)*

**Biology:** It is naked, single-stranded DNA virus with the genome size of 5 kbp and is around 20 – 26 nm in size.

**Disease:** It is the causative agent of the Fifth disease (also called Erythema infectiosum), which is a mild rash illness which usually occurs in children. It appears with bright red cheeks and then the rash spreads onto the extremities, and may be associated with fever and malaise. Acquired immunity protects against further infections.

**Diagnosis:** The diagnosis is made on physical examination and by demonstrating viral antibodies in serology-based assays.

**Treatment:** It is self-limiting.

**Transmission:** The infection is normally transmitted through direct contact with infected individuals.

**Occurrence:** Worldwide.

### *8.1.13 Organism: Hepatitus B virus (HBV)*

**Biology:** HBV is an enveloped, double stranded DNA virus with the genome size of approximately 3 kbp and around 42 nm in size.

**Disease:** HBV causes inflammation of the liver (enlarged liver) with symptoms such as nausea, fatigue, jaundice and abdominal pain but can lead to permanent liver damage or liver cancer.

**Diagnosis:** Hepataitis can be diagnosed by demonstrating the presence of hepatitis B antibodies in patient's blood using serology-based tests.

**Treatment:** Hepatitis B vaccine is the best protection. In addition, the use of interferon and antivirals, lamivudine and adefovir dipivoxil are effective.

**Transmission:** HBV is transmitted to humans via blood and other bodily fluids (unprotected sex, blood transfusions, during birth, etc.).

**Occurrence:** Worldwide (predominantly in developing countries with poor hygiene).

## 8.2 RNA Viruses

### *8.2.1 Organism: Human T cell lymphotropic virus (HTLV)*

**Biology:** These are enveloped viruses, with plus sense single-stranded RNA genome (8 kbp) and is around 80 – 100 nm in size and contain enzyme, reverse transcriptase (RNA-dependent DNA polymerase). The enzyme converts viral RNA genome into the DNA, which is integrated into the host genome. It infects T-lymphocytes, hence the name HTLV. There are two types of viruses, HTLV-I and HTLV-II.

**Disease:** Normally individuals are asymptomatic carriers and do not develop any disease. However, 5% of infected individuals develop the disease after several decades of infection. HTLV-I causes two diseases, 1) Adult T-cell leukaemia (a rare form of cancer in which white blood cells displace normal blood. This leads to infection, shortage of red blood cells (anaemia), bleeding, and other disorders, and often proves fatal) and 2) HTLV-I-associated myelopathy (inflammation of the nerves in the spinal cord characterized by a stiff neck, weakness of legs and backache). In rare cases, HTLV-I may cause inflammation of the eye (uveitis), joints (arthritis), muscles (myositis), lungs (alveolitis) and skin (dermatitis). Diseases due to HTLV-II are not clear but are thought to be CNS-related.

**Diagnosis:** Diseases can be diagnosed by demonstrating HTLV antibodies in blood. Western blotting methods can distinguish between HTLV-I and HTLV-II. Other diagnostic methods include PCR to detect HTLV from the blood.

**Treatment:** No recommended treatment and infection is usually lifelong. Since 95% of infected individuals never develop any disease, treatment remains a low priority.

Leukaemia is usually treated with anti-cancer drugs in combination with antiviral agents.

**Transmission:** HTLV is transmitted via body fluids (blood, semen, vaginal fluids, breast milk) from infected to non-infected individuals, thus preventative measures similar to HIV infection are highly protective.

**Occurrence:** Worldwide (predominantly in Africa, S. America and Asia).

### 8.2.2 *Organism: Human immunodeficiency virus (HIV)*

**Biology:** These are enveloped viruses, with plus sense single-stranded RNA genome (9.2 kbp) and is 100 – 140 nm in size and contain enzyme, reverse transcriptase (RNA-dependent DNA polymerase). The enzyme converts viral RNA genome into DNA, which is integrated into the host genome. It infects cells expressing CD4, mainly macrophages and CD4 T-cells.

There are two types of viruses, HIV-1 and HIV-2.

**Disease:** It involves three stages, 1) primary infection during which the virus enters the human body, replicates and produces flu-like symptoms, 2) asymptomatic infection, during which viral continues to multiply at a lower rate and can lasts for many years (8 to 10 years) and finally 3) symptomatic infection, during which CD4 T-cell count falls below 200 CD4 T-cells per μL of blood and the infected individuals become susceptible to opportunistic pathogens and are characterized as AIDS patients.

**Diagnosis:** The disease can be diagnosed by demonstrating HIV antibodies in the blood using ELISA and can be confirmed by Western blot analysis. Other methods involve the use of PCR to detect viral particles in the blood.

**Treatment:** It includes the use of chemical drugs such as viral enzymes inhibitors, zidovudine (AZT). Other inhibitors of viral enzymes include saquinavir, indinavir, ritonavir and nelfinavir. Now at least in some countries, a combination of antiviral compounds, known as highly active antiretroviral therapy (HAART) is being used. The rationale for this approach is that by combining drugs that are synergistic, non-cross-resistant and no overlapping toxicity, it may be possible to reduce toxicity, improve efficacy and prevent resistance from arising but there are clear side effects and HIV is emerging resistance. It is expected that the vaccine is the ultimate hope in controlling HIV infections.

**Transmission:** HIV is transmitted via body fluids (intimate sexual contact, exposure to contaminated blood via syringes, transfusion etc and perinatal (infected mother to child).

**Occurrence:** Worldwide.

### 8.2.3 *Organism: Poliovirus*

**Biology:** They are naked, plus sense single-stranded RNA viruses with the genome size of approx. 7 kbp and is around 30 nm in diameter.

**Disease:** It causes polio, which is damage of the muscles and nerves which can affect the whole body and can be paralytic.

**Diagnosis:** The diagnosis of polio is based on neurological symptoms and confirmed by using throat or stool samples to demonstrate the presence of viral antigens as it sheds or the presence of viral antibodies in the blood.

**Treatment:** There is no specific treatment but pain killers and moist heat relieves some of the symptoms. The polio vaccine is effective in preventing the disease.

**Transmission:** It is transmitted by contact with infected individuals or with infected secretions (nasal, mouth, faeces), multiplies in the GI tract and spread through the blood.

**Occurrence:** Worldwide

### *8.2.4 Organism: Coxsackievirus*

**Biology:** They are naked, plus sense single-stranded RNA viruses with the genome size of approx. 7 kbp and is around 30 nm in diameter.

**Disease:** It causes head, foot and mouth diseases, which is characterized by rash on the palms and the soles of infected individuals and may cause blisters.

**Diagnosis:** Detection of virus antigens using immunoassays or PCR assays.

**Treatment:** It is self-limiting.

**Transmission:** The disease is transmitted through contact with bodily fluids and by respiratory droplets released into the air by coughing and sneezing.

**Occurrence:** Worldwide.

### *8.2.5 Organism: Echovirus*

**Biology:** They are naked, plus sense single-stranded RNA viruses with the genome size of approx. 7 kbp and is around 30 nm in diameter.

**Disease:** It causes meningitis and meningoencephalitis, which is an inflammation of the membranes covering the brain. It can also cause upper respiratory infections and conjunctivitis.

**Diagnosis:** Detection of virus antigens using PCR assays.

**Treatment:** There is no specific treatment but the use of acyclovir together with anti-seizure drugs, phenytoin and anti-inflammatory drugs, dexamethasone are effective in relieving the symptoms.

**Transmission:** The disease is transmitted through respiratory droplets or through the faecal-oral route.

**Occurrence:** Worldwide

### *8.2.6 Organism: Hepatitus A virus (HAV)*

**Biology:** Hepatitis A virus is a plus sense single-stranded RNA virus (the genome size of 7.5 kbp) which is naked with approx. size of 27 – 30 nm in diameter.

**Disease:** Hepatitis is characterized by fever, malaise, anorexia (persistent loss of appetite), nausea, abdominal pain followed by jaundice (yellowing of the whites of the eye and skin) within a few weeks and may lead to a severely disabling disease within months. It can also prove to be fatal in older patients.

**Diagnosis:** Hepataitis can be diagnosed by demonstrating the presence of hepatitis antibody in the patient's blood using serology-based tests.

**Treatment:** The vaccine is available (two injections 6 to 12 months apart) and recommended for individuals living or travelling to high risk areas (gives protection for around 10 years). Passive immunity may help reduce the severity of the disease. There are no recommended drugs against hepatitis.

**Transmission:** The disease is transmitted from person-to-person by the faecal-oral route by ingestion of contaminated food and water and can also be transmitted by blood transfusions. The virus survives in water and in sewage for long periods of time, thus poor sanitation is a major risk factor in hepatitis.

**Occurrence:** Worldwide.

### *8.2.7 Organism: Rhinovirus*

**Biology:** It is naked, plus sense single-stranded RNA virus (genome size 7.2 kbp) and is around 30 nm in size.

**Disease:** Although it commonly causes colds and rarely causes nasopharyngitis, croup and pneumonia.

**Diagnosis:** Diagnosis is not necessary but can be performed by demonstrating viral antigens in nasal secretions.

**Treatment:** There is no specific treatment, but symptomatic therapy can be helpful.

**Transmission:** Transmitted through the air or close contact with infected persons.

**Occurrence:** Worldwide.

### *8.2.8 Organism: Rubella virus*

**Biology:** It is enveloped, plus sense single-stranded RNA virus (genome size of approx. 9.7 kbp) and is around 70 nm in size.

**Disease:** It causes rubella, which is characterized by fever, headache, malaise, aches, runny nose, rash and very rarely encephalitis.

**Diagnosis:** Diagnosed by demonstrating viral antibodies using immunoassays.

**Treatment:** There is no specific treatment but the use of acetaminophen (for fever) and anti-inflammatory agents can help relieve symptoms. The use of MMR vaccine is effective in preventing the infection.

**Transmission:** Transmitted through the air or close contact with the infected persons or from an infected mother to the foetus, which can result in serious foetal defect.

**Occurrence:** Worldwide.

### *8.2.9 Organism: Eastern equine encephalitis virus*

**Biology:** It is enveloped, plus sense single-stranded RNA virus (genome size of approx. 12 kbp) and is around 70 nm in size.

**Disease:** It causes vector-borne encephalitis, which is inflammation of the brain, leading to coma and death.

**Diagnosis:** Serology-based tests, e.g., ELISA.

**Treatment:** There is no specific treatment but the use of ribavirin together with anti-inflammatory drugs, dexamethasone may be effective in relieving the symptoms.

**Transmission:** It is vector-borne (mosquitoes, *Aedes* spp.).

**Occurrence:** Worldwide (predominant in Africa and Asia).

### *8.2.10 Organism: Yellow fever virus*

**Biology:** It is enveloped, plus sense single-stranded RNA virus (genome size of approx. 10.5 kbp) and is around 40 – 50 nm in size.

**Disease:** It is characterized by high fever, chills, headache, muscle aches, vomiting and backache. After a brief recovery period, the infection can lead to kidney failure, liver failure and shock. Liver failure causes jaundice (yellowing of the skin and the whites of the eyes), hence the disease is called yellow fever.

**Diagnosis:** Serology-based tests, e.g., ELISA.

**Treatment:** There is no specific treatment but the use of ribavirin together with symptomatic therapy may be effective in relieving the symptoms. Vaccine is the best option. People travelling to endemic areas should be vaccinated (vaccine is available).

**Transmission:** It is vector-borne (mosquitoes, *Aedes* spp.). Mosquitoes spread the virus from infected humans to new hosts or infected monkeys to humans.

**Occurrence:** South America and Africa.

### *8.2.11 Organism: Dengue virus*

**Biology:** It is enveloped, plus sense single-stranded RNA virus (genome size of approx. 10.5 kbp) and is around 40 – 50 nm in size.

**Disease:** It is a fever characterized by rash, muscle aches and joint pains, headaches, nausea, vomiting and enlarged lymph nodes and can rarely lead to haemorrhage fever with bleeding from the nose and mouth.

**Diagnosis:** Diagnosed by demonstrating viral antigens or viral antibodies in the serum using ELISA.

**Treatment:** The treatment includes rehydraton together with targetting the fever with acetaminophen.

**Transmission:** It is vector-borne (mosquitoes, *Aedes* spp.). Mosquitoes spread the virus between infected humans to new hosts.

**Occurrence:** Worldwide (except Europe).

### *8.2.12 Organism: West Nile virus*

**Biology:** It is enveloped, plus sense single-stranded RNA virus (genome size of approx. 10.5 kbp) and is around 40 – 50 nm in size.

**Disease:** It is characterized by fever, headache, tiredness, and body aches, which can occasionally develop into encephalitis, meningitis, poliomyelitis with high fever, neck stiffness, disorientation, coma, tremors, muscle aches weakness and paralysis.

**Diagnosis:** Diagnosed by demonstrating viral antibodies in the serum using ELISA.

**Treatment:** The treatment includes rehydraton together with symptomatic therapy.

**Transmission:** It is vector-borne (mosquitoes, *Culex* spp.). Mosquitoes become infected by biting infected birds.

**Occurrence:** Worldwide.

### *8.2.13 Organism: Hepatitis C virus*

**Biology:** It is enveloped, plus sense single-stranded RNA virus (genome size of approx. 10.5 kbp) and is around 40 – 50 nm in size.

**Disease:** It is an inflammation of the liver that may be associated with jaundice, abdominal pain, fatigue, weakness, nausea, vomiting and dark urine. It sometimes causes total liver damage.

**Diagnosis:** Diagnosed by demonstrating viral antigens or viral antibodies in the serum using ELISA.

**Treatment:** There is no recommended treatment for hepatitis C except the use of interferon together with anti-viral, ribavirin, is of value.

**Transmission:** Transmitted via the blood and other bodily fluids (blood transfusions, during birth, etc.).

**Occurrence:** Worldwide.

### *8.2.14 Organism: Severe Acute Respiratory Syndrome (SARS) virus (coronovirus)*

Biology: It is enveloped, plus sense single-stranded RNA virus (genome size of approx. 20 – 30 kbp) and is around 150 nm in size.

**Disease:** Human coronavirus commonly cause colds but a mutation in its genome has resulted in the SARS virus that can cause lungs damage. SARS is a viral pneumonia, with symptoms including fever, dry cough, breathing difficulty, headache and hypoxaemia (low blood oxygen concentration). Death may occur due to progressive respiratory failure due to alveolar damage. The patient may feel better after the first few days of infection, followed by a worsening during the second week, which may be due to the patient's immune responses rather than uncontrolled viral replication.

**Diagnosis:** Diagnosed by demonstrating viral antibodies using serology-based assays or viral particles using PCR assays.

**Treatment:** There is no therapy and the treatment is symptomatic.

**Transmission:** Transmitted via respiratory droplets during sneezing and coughing.

**Occurrence:** Worldwide.

### *8.2.15 Organism: Norwalk virus*

**Biology:** It is naked, plus sense single-stranded RNA virus (genome size of approx. 7.7 kbp) and is around 35 nm in size.

**Disease:** It causes diarrhoeal disease, which is characterized by abdominal pain, diarrhoea, nausea, vomiting and fever.

**Diagnosis:** Normally based on symptoms but PCR assays can be used to detect viral particles.

**Treatment:** There is no recommended treatment against the virus but targetted against the symptoms by excessive intake of fluids (with salts and minerals) and the disease resolves within a few days. Anti-diarrhoeal drugs should not be given.

**Transmission:** Transmitted via contaminated food/water.

**Occurrence:** Worldwide.

### *8.2.16 Organism: Mumps virus*

**Biology:** It is enveloped, minus sense single-stranded RNA virus (genome size of approx. 15 kbp) and is around 200 nm in diameter and 1000 – 10,000 nm long.

**Disease:** It is a painful enlargement or swelling of the salivary glands associated with face pain, fever, headache, sore throat, and swelling of the jaw.

**Diagnosis:** Diagnosed by demonstrating viral antibodies in serum using ELISA.

**Treatment:** There is no specific treatment but gargling with warm, salted water and corticosteroid relieves the pain. The use of MMR vaccine is effective in preventing the infection.

**Transmission:** Transmitted via respiratory droplets from infected persons.

**Occurrence:** Worldwide.

### *8.2.17 Organism: Measles virus*

**Biology:** It is enveloped, minus sense single-stranded RNA virus (genome size of approx. 15 kbp) and is around 200 nm in diameter and 1000 – 10,000 nm long.

**Disease:** It is characterized by fever, a cough, runny nose, fever, a sore throat, inflammation of eyes, photophobia, and rash within a few days of contracting the virus.

**Diagnosis:** Diagnosed by demonstrating viral antibodies using ELISA.

**Treatment:** There is no specific treatment but the use of MMR vaccine (measles, mumps and rubella) is effective in preventing this infection.

**Transmission:** Transmitted via respiratory droplets from infected persons.

**Occurrence:** Worldwide.

### 8.2.18 *Organism: Respiratory syncytial virus*

**Biology:** It is enveloped, minus sense single-stranded RNA virus (genome size of approx. 15 kbp) and is around 200 nm in diameter and 1000 – 10,000 nm long.

**Disease:** It causes bronchiolitis, inflammation of the bronchioles (small air way branches in the lungs) characterized by short, rapid breathing with a cough.

**Diagnosis:** Diagnosed by demonstrating viral antigens or viral antibodies using ELISA.

**Treatment:** There is no specific treatment (usually self-limiting), but oxygen therapy and dexamethasone or antiviral agent, ribavirin may be effective.

**Transmission:** Transmitted via respiratory droplets from infected persons.

**Occurrence:** Worldwide.

### 8.2.19 *Organism: Human parainfluenza virus-1*

**Biology:** It is enveloped, minus sense single-stranded RNA virus (genome size of approx. 15 kbp) and is around 200 nm in diameter and 1000 – 10,000 nm long.

**Disease:** It causes bronchiolitis, inflammation of the bronchioles characterized by short, rapid breathing with a cough.

**Diagnosis:** Diagnosed by demonstrating viral antigens or viral antibodies using ELISA.

**Treatment:** There is no specific treatment, but oxygen therapy and dexamethasone or antiviral agent, ribavirin may be effective.

**Transmission:** Transmitted via respiratory droplets from infected persons.

**Occurrence:** Worldwide.

### 8.2.20 *Organism: Rabies virus*

**Biology:** It is enveloped, minus sense single-stranded RNA virus (genome size of 12 kbp) and is around 180 nm long and 75 nm in diameter.

**Disease:** It causes acute encephalitis leading almost always to death. Following an initial incubation period (a few days to a few months), the virus replicates in the muscle cells and enters into the peripheral nerves and is transported to the CNS via the sensory and motor nerves exhibiting behavioural changes. Initially, flu-like symptoms are observed together with headache, fever, discomfort followed by anxiety, confusion, agitation, abnormal behaviour associated with encephalitis leading almost always to death. Around 70,000 deaths are caused annually.

Almost all mammals can be infected by this virus and few act as virus reservoirs (foxes, dogs, raccoons, bats, etc.).

**Diagnosis:** Most commonly used assays are PCR (for saliva samples) and immunoassays (for serum and cerebrospinal fluids),

**Treatment:** Although there is no treatment, post-prophylactic vaccine is useful, i.e., rabies vaccine which provides immunity to rabies when administered early after an exposure to the virus.

**Transmission:** The disease initiates when infected saliva of a host is passed to an uninfected host (usually through a bite). Other modes of transmission are various routes of transmission which have been documented and include contamination of mucous membranes (i.e., eyes, nose, mouth), aerosol transmission and corneal transplantations. The most common mode of the rabies virus transmission is through a bite of an infected animal.

**Occurrence:** Worldwide (but not in the British Isles, some other parts of Western Europe and Australasia)

### *8.2.21 Organism: Ebola virus*

Biology: It is enveloped, minus sense single-stranded RNA virus (genome size of 19 kbp) and is around 80 nm in diameter and up to 800 nm in length.

**Disease:** It causes haemorrhagic fever, characterized by fever, headache, rash, fatigue, malaise, backache, vomiting, bleeding abnormalities, shock and death (up to 90% mortality). The symptoms appear within one week of contracting the virus.

**Diagnosis:** Demonstration of viral antigens and viral antibodies using serology-based assays.

**Treatment:** There is no treatment, but symptomatic therapy is helpful.

**Transmission:** Transmitted to humans via contact with infected animals or their products.

**Occurrence:** Africa.

### *8.2.22 Organism: Lassa fever virus*

**Biology:** It is enveloped, minus sense single-stranded RNA virus (genome size of 10 – 11 kbp) and is around 100 – 200 nm in size.

**Disease:** It causes haemorrhagic fever, characterized by fever, headache, rash, fatigue, malaise, backache, vomiting, bleeding abnormalities, shock and death (up to 90% mortality). The symptoms appear within one week of contracting the virus.

**Diagnosis:** Demonstration of viral antigens and viral antibodies using serology-based assays.

**Treatment:** The treatment includes supportive care together with the use of antiviral, ribavirin, especially during the early course of infection.

**Transmission:** Transmitted to humans via contact with rodents or their faeces.

**Occurrence:** Africa.

### *8.2.23 Organism: Influenza virus (three types, A, B, C)*

Biology: These are enveloped, minus sense single-stranded RNA (genome size is 13.6 kbp) and is around 15nm in diameter and 200 – 3000 nm long. Type A can cause infections both in humans and animals while type B and C only cause infections in humans. Both A and B type viruses are common causes of epidemic every winter. Type C virus causes a mild respiratory infection (flu shot protects against A and B but not against type C influenza). Type A virus is further divided into subtypes based on two proteins on their surface called hemagglutinin (H) and neuraminidase (N). The current subtypes of influenza. A viruses found in people are influenza virus A(H1N1 and H3N2). Type B virus is not divided into subtypes. Influenza A(H1N1 and H3N2), and influenza B strains are included in each year's influenza vaccine.

**Disease:** Influenza (flu) is an infection of the respiratory tract (different from the common cold which is a mild infection caused by a variety of viruses). Flu is caused by the influenza virus and is characterized by fever, chills, muscle aches and pains, sweating, coughing, nasal congestion, sore throat, headache and fatigue killing more than 25,000 people per year in the US alone (around 5,000 in the UK) in the non-epidemic years.

**Diagnosis:** The virus is cultured from deep nasal swabs or the throat. Serological tests using blood indicates antibody levels and confirms the infection or by indirect immunofluorescence to demonstrate the presence of viral antigens. But usually, diagnosis is based on symptoms.

**Treatment:** The influenza vaccine (flu shot) is a good prophylactic measure especially for individuals who are over 65 or suffer from chronic diseases. The vaccine kills the influenza virus which is given every year. This is due to the limited protection of the vaccine as well as changes in the virus ( the vaccine is usually the most likely viral strain at the time). Chemotherapeutic approaches include antiviral compounds such as amantidine, which inhibits the growth of the influenza virus A by interfering with the uncoating of the virus and can also act prophylactically.

Other drugs include rimantadine which can be used in place of amantadine for prophylaxis and treatment. In addition, Zanamivir, a neuroaminidase inhibitor is effective against both influenza A and B and administered by inhalation only for treatment and not prophylaxis. Other neuroaminidase inhibitors are oseltamivir, which is given orally both for treatment and prophylaxis.

**Transmission:** Type A is found in many animals such as ducks, chickens, pigs, horses, etc. as well as humans while wild birds act as natural reservoirs. Type B transmits between humans. Certain strains (e.g., H5, H7) can cause widespread death in birds such as chickens while other strains have limited effects on birds. Pigs can be infected with swine influenza viruses as well as human and avian influenza viruses. This may allow viruses to mix and create a new virus, which may have genes from the human virus but the surface protein (i.e., H and N) from the bird virus. This virus will be able to infect humans and spread between people (via infected droplets which are sneezed or coughed into the air) but have different surface proteins and individuals will not have any immunity against such a virus and this could result in pandemic.

**Occurrence:** Worldwide.

### *8.2.24 Organism: Hantavirus*

**Biology:** Hantaviruses are enveloped, minus sense single-stranded RNA viruses (genome size of 11.8 – 13.8 kbp) and is around 90 – 100 nm in diameter.

**Disease:** It causes respiratory disease syndrome (hantavirus pulmonary syndrome) and is characterized by fever, chills, muscle aches and shortness of breath, nausea, vomiting, resulting in respiratory failure leading to death (up to 50% mortality rate).

**Diagnosis:** Serology-based tests for hanta virus antigens are used to confirm this infection. Other common symptoms include hypoxia (decreased saturation of oxygen in the blood), hypotension (low blood pressure) and respiratory distress.

**Treatment:** No recommended treatment or vaccine is available for this infection. Currently, oxygen therapy is used in the intensive care units and closely monitored by blood gases. The use of ribavirin is currently being tested in experimental models.

**Transmission:** The disease spreads via infected deer mice. Infected rodents shed the virus through urine, droppings, and saliva. The virus is transmitted to humans by breathing in contaminated material. It is not transmitted from person-to-person or from farm animals such as dogs, cats or pet rodents.

**Occurrence:** Worldwide.

#### *8.2.25 Organism: Rotavirus*

**Biology:** Wheel-like appearance, hence the name (*rota* is a Latin word meaning wheel) and is 60 – 80 nm in size. They are environmentally stable, non-enveloped, double-stranded RNA viruses (genome size of 16 – 21 kbp).

**Disease:** Rotavirus causes severe diarrhoea (especially in infants and young children) causing nearly 1 million annual deaths. Following a few days of incubation, the disease is produced and is characterized by vomiting, fever, abdominal pain and watery diarrhoea for up to two weeks.

**Diagnosis:** Detection of rotavirus antigens in stools using immunoassays or PCR assays.

**Treatment:** This is a self-limiting infection but requires excessive intake of fluids to prevent dehydration.

**Transmission:** Transmission is normally via the faecal-oral route. Since the virus is stable in the environment, transmission can occur through ingestion of contaminated water/food.

**Occurrence:** Worldwide.

## 9. HUMAN IMMUNODEFICIENCY VIRUS (HIV) AS THE MODEL VIRION

### 9.1 Taxonomy and Characteristics

HIV belongs to the retrovirus family *Retroviridae*. There are three sub-families:

1. Deltaretrovirus, which includes human T-lymphotropic virus type-1 (HTLV-1) causing T-cell leukaemia.
2. Spumavirus, which include human foamy viruses (HFV – no disease is known).
3. Lentivirus, which include human immunodeficiency virus type-1 [HIV-1, a major causative agent of acquired immune deficiency syndrome (AIDS)] and HIV-2, a causative agent of slowly progressing AIDS.

## 9.2 Common Features of Lentiviruses

- Lentiviruses can infect non-dividing cells such as macrophages.
- Lentivirus genomes are integrated into the host genome.
- Lentiviruses have high mutation rates.
- The genomes of lentiviruses encode regulatory proteins, which regulate viral transcription.
- They possess major human pathogens, HIV-1, HIV-2.
- Their size varies from 80 – 130 nm with icosahedral symmetry.
- They have an envelope and a genome size of approx. 10 kb.
- Their genome is replicated in the nucleus and virions are assembled in the cytoplasm.
- They produce chronic infections such as AIDS.

## 9.3 Discovery and Origin of HIV

The discovery of HIV followed the discovery of the disease, i.e., AIDS. In 1981, several cases of *Pneumocystis carinii* pneumonia (PCP) were observed in Los Angeles, USA. This was highly unusual as even a single case of PCP was uncommon. This cluster of PCP infections suggested something unusual. Subsequently in 1983, Francoise Barre-Sinoussi, Claude Chermann and Luc Montagnier at the Pasteur Institute, France isolated a retrovirus from the lymph node cells of a patient of lymphadenopathy (swollen lymph glands) and called it lymphadenopathy-associated virus. The following year, Robert Gallo at the National Institute of Health, USA confirmed this finding, linking the virus to the immunodeficiency syndrome, AIDS. In 1986, a second HIV was isolated from West Africa (HIV-2). The evolutionary and phylogenetic studies suggest that HIV originated via a host shift from chimpanzees to humans involving blood products or from unidentified monkeys in central Africa around 1930s.

## 9.4 AIDS Epidemic

Since its first appearance in 1981, HIV has become a global epidemic (Table 1). So far more than 20 million people have died of HIV with more than 40 million people infected with HIV as of 2005. The most distressing aspect is that more than 80% of infections are in the developing countries with no or limited antiretroviral therapy. The virus is spreading, evolving and mutating at an alarming rate. For example, there were at least 5 million new reported cases of HIV/AIDS in 2003 alone (approximately 14,000 infections per day), while 3 million deaths occurred due to HIV/AIDS-related diseases (approximately 8,500 deaths per day) resulting in millions

**Table 1** Regional distribution of people with HIV as of 2005 (source WHO).

| *Region* | *Infected people living with HIV* | *Newly infected* | *Deaths* |
|---|---|---|---|
| Sub-Saharan Africa | | | |
| 2005 | 25.8 million | 3.2 million | 2.4 million |
| 2003 | 24.9 million | 3 million | 2.1 million |
| North Africa and Middle East | | | |
| 2005 | 510,000 | 67,000 | 58,000 |
| 2003 | 500,000 | 62,000 | 55,000 |
| South and South East Asia | | | |
| 2005 | 7.4 million | 990,000 | 480,000 |
| 2003 | 6.5 million | 840,000 | 390,000 |
| East Asia | | | |
| 2005 | 870,000 | 140,000 | 41,000 |
| 2003 | 690,000 | 100,000 | 22,000 |
| Oceania | | | |
| 2005 | 74,000 | 8,200 | 3,600 |
| 2003 | 63,000 | 8,900 | 2,000 |
| Latin America | | | |
| 2005 | 1.8 million | 200,000 | 66,000 |
| 2003 | 1.6 million | 170,000 | 59,000 |
| Caribbean | | | |
| 2005 | 300,000 | 30,000 | 24,000 |
| 2003 | 300,000 | 29,000 | 24,000 |
| Eastern Europe and Central Asia | | | |
| 2005 | 1.6 million | 270,000 | 62,000 |
| 2003 | 1.2 million | 270,000 | 36,000 |
| Western and Central Europe | | | |
| 2005 | 720,000 | 22,000 | 12,000 |
| 2003 | 700,000 | 20,000 | 12,000 |
| North America | | | |
| 2005 | 1.2 million | 43,000 | 18,000 |
| 2003 | 1.1 million | 43,000 | 18,000 |
| Total | | | |
| 2005 | 40.3 million | 4.9 million | 3.1 million |
| 2003 | 37.5 million | 4.6 million | 2.8 million |

of orphans worldwide. Thus, there is a clear and urgent need for the HIV vaccine or the development of cheap, affordable antiretroviral chemotherapy.

## 9.5 Natural History of HIV Infection

The natural history of HIV infection can be divided into three stages which are as follows.

### 9.5.1 *Primary infection*

During this stage, the virus enters the human body, replicates and produces viremia and may result in symptoms such as mononucleosis within a few weeks together with fever, enlarged lymph nodes, sore throat, muscle aches, rash and tiredness lasting less than a month (influenza-like symptoms). Viral antibodies can only be detected during the acute illness. The immune system of the body responds aggressively and introduces cytotoxic T-cells that kill the majority of viral-infected cells (CD4 T-cells), thus bringing the virus under control to around $10^3 - 10^5$ copies of viral RNA per mL of blood. In the process, CD4 T-cells are reduced but their level becomes relatively normal after the initial acute phase of the disease.

### 9.5.2 *Asymptomatic infection (latent phase)*

This is a carrier stage which follows initial infection and can lasts for many years (8 to 10 years) before the development of the clinical disease. However, the virus replicates continuously during this stage, and viral particles can be detected in the blood (but at a much lower level, viral RNA in the blood is around $10^3$ copies per mL) and thus no symptoms are observed. Regardless of the low levels of viral particles, the most significant feature of this stage is the gradual decline in the number of circulating CD4 T-cells. It is the loss of CD4 T-cells which makes an individual susceptible to opportunistic pathogens that characterize AIDS. In some cases, infection does not proceed beyond this asymptomatic phase and CD4 counts remain stable, i.e., 1,000 CD4 T-cells per μL of blood (these people are known as long-term survivors). The opportunistic pathogens can not produce infections in such adults. However the continued decline of these cells brings their levels to below 200 CD4 T-cells per μL of blood and such individuals become susceptible to opportunistic pathogens and are characterized as AIDS patients.

### 9.5.3 *Symptomatic HIV infection and AIDS*

Once the CD4 T-cell count falls below 200 CD4 T-cells per μL of blood, the infected individuals become susceptible to opportunistic pathogens and are characterized as AIDS patients. During this stage, the virus produces rapidly and the level of infectious viral RNA in the blood reaches around $10^6 - 10^8$ copies per mL. The host immune system responds by massive production of CD4 T-cells to replace massive killing of CD4 T-cells. With the requirement to replace billions of T-cells per day, the body's immune system is worn out and is no longer able to replace CD4 T-cells. In addition, the continuous evolution of the HIV-1 genome overwhelms the immune system's ability to respond and results in the rapid progression of the disease, i.e., the patients become prone to a variety of opportunistic pathogens. The susceptibility of individuals to the opportunistic diseases is used as a measure to define AIDS (Table 2). Overall, HIV infection has at least three stages:

1. Initial infection with rapid viral replication and dissemination, often accompanied by a transient period of disease.
2. A latent period during which the virus is brought under immune control and no disease occurs.
3. Finally high levels of viral replication resumes later resulting in the disease.

**Table 2** AIDS-defining illnesses. These include infections with organisms of low virulence that usually do not cause infections in immunocompetent hosts (source; centres for disease control).

| *Infection or cancer 1982* | *Site* |
|---|---|
| Cryptosporidiosis | Diarrhoea for >1 month |
| *Pneumocystis carinii* | Lungs |
| Candidiasis | Oesophagus |
| Cryptococcosis | CNS or disseminated |
| Toxoplasmosis | Other than liver, spleen, lymph nodes |
| *Mycobacterium avium* | Other than lungs, lymph nodes |
| *Mycobacterium kansasii* | Other than lungs, lymph nodes |
| Cytomegalovirus | Other than liver, spleen, lymph nodes |
| Herpes simplex virus | Lungs, gastrointestinal tract |
| Progressive multifocal leucoencephalopathy | Brain |
| Kaposi's sarcoma | All |
| Lymphoma | Brain |
| *1985 – The above disease plus the following with laboratory evidence of HIV infection* | |
| Histoplasmosis | Disseminated |
| Candidiasis | Bronchi, lungs |
| Isosporiasis | Gastrointestinal tract |
| NonHodgkin lymphoma | All |
| Lymphoid interstitial pneumonitis | Lungs |
| *1987 – The above disease plus the following with laboratory evidence of HIV infection* | |
| Multiple pyogenic bacteria | All (in children) |
| Coccidioidomcosis | All |
| HIV encephalopathy | Brain |
| Mycobacterium tuberculosis | Extrapulmonary |
| *Salmonella* bacteremia | Blood |
| Mycobacteriosis | Disseminted |
| *1993 – The above disease plus the following with laboratory evidence of HIV infection* | |
| *Mycobacterium tuberculosis* | Lungs |
| Recurrent bacterial pneumonia | Lungs |
| Invasive cervical cancer | |

## 9.6 Modes of Transmission

The mechanisms of HIV transmission include the transfer of body fluids from an infected individual to a new host and can be broadly categorized as follows:

1. Intimate sexual contact with an infected individual.
2. Exposure to infected blood, e.g., through blood transfusions (although all blood samples from blood donors are now screened for HIV) and through the use of contaminated needles (especially in the intravenous drug users).
3. Perinatal transmission, this involves transmission from infected mothers to their infants and can occur by i) uterus to unborn child, but does not always occur and some children are born to infected mothers remain HIV negative or through ii) breast feeding.

However, HIV is not transmitted by casual contact such as a hand shake, casual kissing, sneezing, as HIV does not spread through air (unlike viruses that cause upper respiratory infections) and require bodily fluid transfer.

## 9.7 Diagnosis

Diagnosis is made by demonstration of HIV antibodies (in particular to Env protein) in the blood using an ELISA. However, to eliminate the risk of false negative, Western blot analysis is used to confirm the presence of HIV antibodies. Other recent methods involve the use of PCR to detect viral particles in the blood. PCR detects genomic RNA of the virus particle. This method requires reverse transcription of viral RNA into the DNA, followed by its detection using PCR and is called RT-PCR. This is a useful method as it can detect the infection before the production of antibodies as well as monitor the status of the disease, i.e., viral load during infection and/or during the antiretroviral treatment. However, in many countries HIV testing is not available and alternative approaches have been recommended (Fig. 7).

## 9.8 HIV Treatment

At present, HIV infection is treated using chemical drugs. These include viral enzymes inhibitors such as zidovudine (3'-azido-2',3'-dideoxythymidine or AZT) which is a potent reverse transcriptase inhibitor. This is a nucleoside and when taken up by host cells is phosphorylated to a triphosphate form and resembles deoxythymidine triphosphate (dTTP), a building block of DNA. However, the host cell DNA polymerases do not use AZT but the viral reverse transcriptase enzyme incorporates AZT into

**A) WHO definition for AIDS, where HIV testing is available.**

For the purpose of AIDS surveillance a person is considered to have AIDS, if a test for HIV antibody gives a positive result, and one or more of the following conditions are present:

1. More than 10% bodyweight loss with diarrhoea or fever, or both, intermittent or constant, for at least 1 month, not known to be due to a condition unrelated to HIV infection.
2. Cryptococcal meningitis.
3. Pulmonary or extrapulmonary tuberculosis.
4. Kaposi's sarcoma.
5. Neurological impairment that prevent daily activities, not known to be due to a condition unrelated to HIV infection.
6. Candidiasis of the oesophagus (which may be presumptively diagnozed based on the presence of oral candidiasis accompanied by dysphagia).
7. Clinically diagnozed life-threatening or recurrent episodes of pneumonia, with or without aetiological confirmation.
8. Invasive cervical cancer.

**B) WHO case definition for AIDS, where HIV testing is not available**

For the purposes of AIDS surveillance, a person is considered to have AIDS if two of major signs are present together with one of minor signs, and if these signs are not known to be due to a condition unrelated to HIV infection

Major signs

- Weight loss of more than 10% bodyweight
- Chronic diarrhoea for more than one month
- Prolonged fever for more than one month (intermittent or constant)

Minor signs

- Persistent cough for more than one month
- Generalized dermatitis
- Oropharyngeal candidiasis
- Chronic progressive or disseminated herpes simplex infection
- Generalized lymphadenopathy

The presence of either generalized Kaposi's sarcoma or cryptococcal meningitis is sufficient for the diagnosis of AIDS for surveillance purposes.

**Fig. 7 A)** WHO definition for AIDS, where HIV testing is available and **B)** WHO case definition for AIDS, where HIV testing is not available.

growing viral DNA during reverse transcription. Since AZT lacks 3'-OH group, it cannot serve as a site for the addition of the next nucleotide and viral DNA chain is terminated. However, recently other inhibitors of viral reverse trancriptase as well as viral protease inhibitors such as saquinavir, indinavir, ritonavir and nelfinavir) have been developed. Viral protease is crucial for the proteolytic cleavage of viral proteins into functional units as described below. Other compounds in the pipeline are the CCR5 and

CXCR4 antagonists that block viral interactions with the host cells as well as inhibitors of viral enzyme, integrase. A major problem in the inability of these compounds to clear the infection is due to the ability of the virus to rapidly mutate and evade the drug killing, thus a combination of drugs against multiple targets is more effective. Now at least in some countries, a combination of antiviral compounds, known as highly active antiretroviral therapy (HAART) are being used. The rationale for this approach is that by combining drugs that are synergistic, non-cross-resistant and no overlapping toxicity, it may be possible to reduce toxicity, improve efficacy and prevent resistance from arising. Although HAART is highly effective and delays the disease but HIV has already shown the ability to persist, mutate and evade the therapy. It is expected that the vaccine is the ultimate hope in controlling HIV and several targets such as gp120, alone or in combination of other HIV proteins are being tested in clinical trials.

## 9.9 Types of HIV-1

HIV-1 can bind to immune cells expressing CD4, in particular macrophages and T-cells. In addition, HIV-1 makes use of a co-receptor, i.e., chemokine receptor CCR5 or CXCR4. Binding to the co-receptor is dependent on V3 region (variable loop). Now we know that the chemokine receptor plays a crucial role in HIV entry into the host cells. Chemokine receptors are 7-transmembrane G-coupled proteins. Based on the receptor binding there are the following types of HIV-1:

1. T-tropic strains of HIV-1 bind to CD4 + CXCR4. These viruses replicate well in T-cells and are called T-tropic.
2. M-tropic strains of HIV-1 bind to CD4 + CCR5. These viruses replicate well in macrophages and are called M-tropic.
3. Dual-tropic strains of HIV-1 bind to CD4 + CXCR4 or CCR5. These viruses can replicate in both cell types.

Of interest, both macrophages and T-cells have CD4, while CCR5 is predominantly expressed on macrophages and CXCR4 is expressed on T-cells.

## 9.10 HIV Biology: Proteins and their Role in the Pathogenesis

The genomes of HIV are single stranded RNA genomes with typical genome size of 9,200 ribonucleotide subunits or bases coding for about 10 proteins as described below (as compared to humans which is composed of approximately 3,200,000,000 base pairs coding for approx. 30,000 proteins). HIV-1 has following major genes:

(1) *gag* – encoding structural proteins (matrix, capsid and nucleocapsid).
(2) *pol* – encoding polymerase (reverse transcriptase), integrase and protease.
(3) *env* – encoding coat protein.
(4) *tat, rev, nef, vpr, vpu* and *vif* – encoding proteins that regulate various aspects of the virus life cycle.

### 9.10.1 Env protein (required for binding to the host cells)

This is an outer coat protein (envelope protein), is initially synthesized as gp160 (160 KD glycoprotein composed of 895 aa). During the transport to plasma membrane through Golgi apparatus, it is cleaved into gp120 (550 aa) and gp41 (345 aa containing the transmembrane domain) by cellular proteases but both units remain attached non-covalently. Gp120 is then expressed on the outer surface of the virion that binds to the host cells. Gp120 is heavily glycosylated with carbohydrate accounting for upto 40-50% of its molecular weight and considered as a silent face of gp120 as it plays a role in immune evasion. Gp120 consists of five constant regions (C1-C5) interspersed with five variable regions (V1-V5). The initial binding to the host cell receptor repositions V1 and V2 and exposes V3 loop (co-receptor binding region).

### 9.10.2 Gag

The *gag* gene gives rise to structural proteins, matrix (MA), capsid (CA), nucleocapsid (NC) and p6 that are crucial for various viral functions which are as follows:

(1) MA is required for targetting Gag to the plasma membrane prior to viral assembly.
(2) CA is important in the assembly of virus core.
(3) NC binds to the full-length viral RNA and delivers it into the assembling virion.

The viral protease cuts apart viral proteins into function functional units. For example, the HIV-1 *gag* gene contains the genetic information for the matrix, capsid and nucleocapsid proteins and is initially translated into one long protein containing all three of these proteins linked together. It is the viral protease that cuts them apart so that they can carry out their individual functions.

In addition to the *gag, pol* and *env* genes, HIV-1 has six other 'accessory' genes that encode proteins which regulate various aspects of the virus life cycle. These are *tat, rev, nef, vpr, vpu* and *vif*; *tat* and *rev* regulate the expression of viral genes at the transcriptional and post-transcriptional levels, respectively; *nef* and *vpu* affect host cell expression of CD4; *vpr* can cause cell cycle arrest; the functions of *vif* are not well understood.

### 9.10.3 *Pol*

The *pol* gene encodes the following enzymes:

1. reverse transcriptase (RT) which carries out the reverse transcription process.
2. Integrase (IN) that mediates the insertion of the reverse transcribed viral DNA into the host genome.
3. Protease (PR) that cleave viral proteins to produce functional units (Gag into MA, CA, NC, p6 and Pol into RT, IN, PR).

### 9.10.4 *Tat*

This is a potent trans-activator of transcription, accelerating viral protein production several thousand times (in particular Rev, Nef and itself). In the absence of Tat, HIV transcription initiates for few hundred ribonucleotides but fail to transcribe the entire 9,200 bases viral genome. This is particularly important because most HIV-1 mRNA transcripts terminate prematurely in the absence of Tat.

### 9.10.5 *Vpu*

The newly synthesized Env are sometimes held in the ER through interactions with newly synthesized CD4. Vpu binds to CD4 and promotes its CD4 degradation, allowing Env transport to the cell surface for viral assembly. Variations on *vpu* gene results in decreased release of virions thus it is required for the smooth release of virions from the infected cells.

### 9.10.6 *Nef*

Nef is suggested to reduce levels of CD4 by facilitating routing of CD4 from the cell surface to lysosomes and from newly synthesized CD4 from Golgi to lysosomes promoting virion release. It is also suggests promoting downregulation of MHC-I molecules thus protecting the killing of the infected cells by the cytotoxic T-cells. Nef mutants exhibit decreased rate of viral genome synthesis following infection. It also negatively regulates Tat and Rev and inhibits induction of NF-kB thus promoting the latency period.

### 9.10.7 *Vpr*

Following entry, the virus is uncoated and nucleoprotein complexes are transported to the host cell nucleus by Vpr protein. Thus, it contains a NLS (nuclear localization signal) that directs transport to the nucleus.

### 9.10.8 *Vif*

Vif plays a role in the maturation of infectious virion and increases its infectivity, thus increased transmission. The *vif*-mutant shows altered core structure.

### 9.10.9 *Rev*

Rev is important in the transport of viral mRNA from the nucleus into the cytoplasm (shuttles between the nucleus and the cytoplasm). It promotes export of unspliced RNA (which encodes viral structural proteins) from the nucleus to the cytoplasm. More Rev means greater amount of expression of Env proteins. In the absence of Rev, late HIV-1 mRNAs do not exit nucleus.

### 9.10.10 *Tnv/Tev*

A fusion protein resulting from a frameshift reading the mRNA (mixes up Tat/Env/Rev), and has full functional capacity of both Tat and Rev.

## 9.11 HIV Replication Steps

These can be divided into the following steps.

### 9.11.1 *Attachment to CD4 cells*

HIV expressing gp120 on the outer coat have high affinity for CD4 expressing cells, which allows its attachment to T-cells and macrophages. However, this is not sufficient to allow virus entry into the immune cell. This concept was established with the discovery of the chemokine receptors (CCR5 /CXCR4) that act as co-receptors for gp120. The initial binding of gp120 and CD4 triggers the structural changes in gp120 exposing its V3 loop. The region of gp120 including the V3 loop (bridging sheet) domain binds to the chemokine receptor thus triggering the actual entry of HIV into the host cells.

### 9.11.2 *Viral DNA synthesis*

The binding of gp120 with CD4+ CCR5/CXCR4 exposes N-terminus gp41 of the viral protein pulling both viral and host membranes together causing their fusion. This allows the entry of the viral genetic material into the cell. Virion core is uncoated and exposes viral nucleoprotein complex, which among other proteins contains viral genomic RNA. First, viral RNA is converted into double-stranded DNA by viral enzyme, reverse transcriptase. It is noteworthy that the viral polymerase (unlike cellular DNA polymerases) do not proofread the sequence and thus exhibit a higher tendency of mutation rate, estimated error rate is about $3 \times 10^{-5}$ errors per base per replication cycle compared with $10^{-9}$ errors per base in the host. This enables the virus to constantly evolve and presents a major problem for the host immune system and for the antiretroviral therapy. The process of reverse transcription occurs in a complex, called pre-integration complex in the cytoplasm. This complex is then transported to the nucleus and viral DNA is incorporated into the host genome.

### 9.11.3 *Viral production*

Next, the complex containing viral DNA is transported into the nucleus and inserted into the DNA chromosome at random sites with the aid of viral integrase. This involves covalent bonding of viral DNA to the host cell chromosomal DNA. Viral genes are transcribed into the viral RNA using host cell transcriptional machinery. These RNA are either used as mRNA to form viral proteins as well as incorporated into the virion. Early steps include the synthesis of the regulatory protein, Tat, Rev, Nef that set the scene for rapid viral production. Viral structural proteins and enzymes are synthesized as large polyprotein precursor, Gag-Pro-Pol that is packed into the virion as precursor protein. In contrast, viral Env protein is initially synthesized as gp160 proteins and is cleaved into gp120 and gp41 by the host cellular protease in the Golgi apparatus during its translocation to the host plasma membrane.

### 9.11.4 *Assembly and release*

Once the proteins and other viral components are made (i.e., genomic RNA, Gag and Gag-Pro-Pol precursor proteins), virions are assembled at the host cell plasma membrane (already contains gp120/gp41 complex), which then buds out from the surface of the infected cell using a process called 'budding'. In the process, the virion encloses itself in the plasma membrane which acts as viral envelope, that contains gp120 which bind to the next CD4 expressing cells, thus completing the cycle. Once released from the cell, immature, non-infectious virions undergo proteolytic cleavage to become mature, active virions.

CHAPTER

4

# Bacteria

## 1. INTRODUCTION

Traditionally, micro-organisms are differentiated based on their properties and divided into intracellular parasites (viruses), prokaryotes (Latin 'pro' means before and Greek 'karyon' means nucleus; without membrane-enclosed nucleus) and eukaryotes containing a membrane-enclosed nucleus (fungi and protozoa). However, the development of molecular methods in recent years divided all living organisms into three domains (by Carl Woese in 1981 based on ribosomal RNA sequences).

(1) The ancient 'Archaea' as in Archaea era (3.9 – 2.6 billion years ago) are prokaryotes which inhabit the extreme environments, i.e., halophiles, methanogenic, thermophiles. The first cells were most likely the anaerobes due to the absence of oxygen in the atmosphere.

(2) Bacteria domain are prokaryotes that consists of an extensive group of microorganisms responsible both for our existence i.e., so called 'good bacteria' and some of which are known to cause the most devastating human infections. Of note, rocks as old as 3.5 billion years old showed prokaryote fossils.

(3) Eukarya, including all remaining nucleated uni- and multicellular organisms. Within eukaryotes, protists evolved approximately 1.5 billion years ago (diplomonads e.g., *Giardia* is thought to be among the ancient single-cell eukaryotes which possesses nucleus but lacks mitochondria), leading to the emergence of remaining single-cell organisms such as amoebae and higher animals.

Bacteria are single-celled organisms which belong to prokaryotes, i.e., their chromosomal DNA is not enclosed in a nuclear membrane (lack nucleus). They also lack Golgi apparatus (organelles associated with glycosylation), mitochondria (organelles associated with energy production) and endoplasmic reticulum (Fig. 1). Bacteria can utilize a

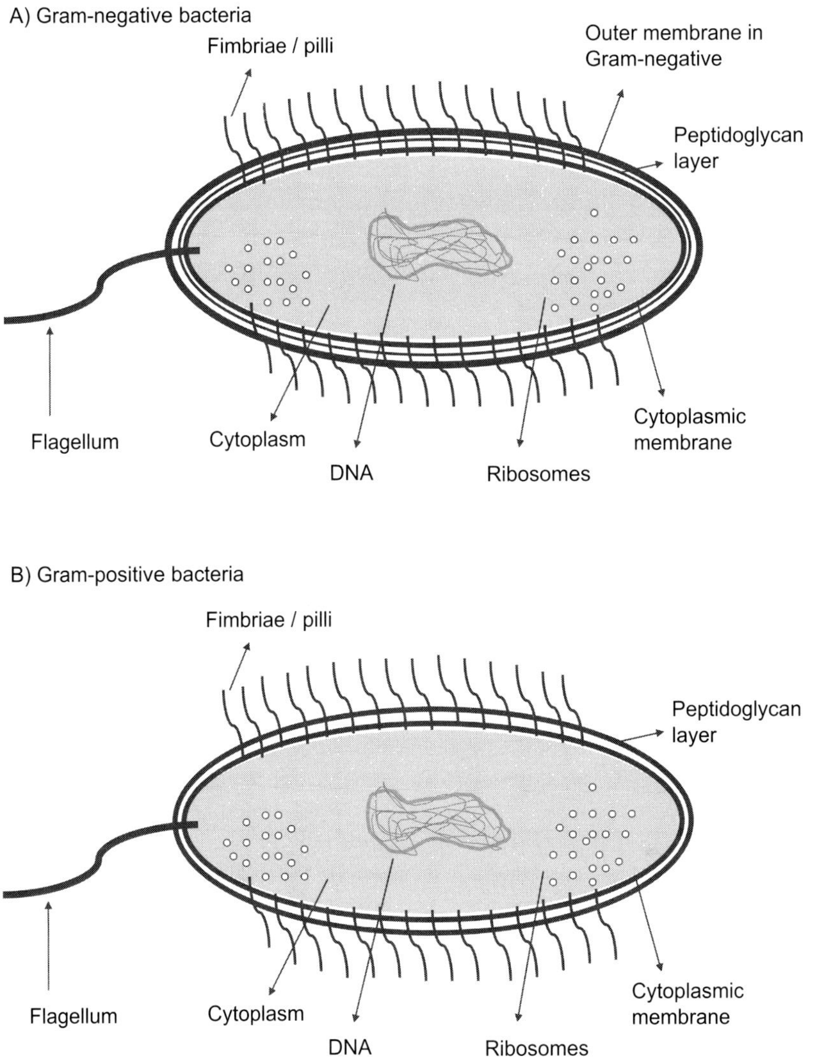


**Fig. 1** **(A)** Gram-negative bacteria and **(B)** Gram-positive bacteria. Cell wall in Gram-negative consists of an outer membrane and a peptidoglycan layer. In contrast, Gram-positive bacteria lack outer membrane and only contain a peptidoglycan layer. The cell wall in Gram-positive consists of several peptidoglycan layers forming a thick wall.

diverse array of materials as a food source including hydrocarbon substrates, also phenol, rubber, petroleum, etc., and therefore can colonize almost every niche on our planet. As described above, following the advances of molecular tools and the discovery of DNA sequencing, bacteria were divided into two groups, archaebacteria (i.e., ancient bacteria) and eubacteria (true bacteria). The former includes bacteria that have different cell walls and membranes and can exist in extreme environments such as hot springs (*Sulfolobus* spp. and *Pyrococcus* spp.) and high salinity (*Halobacterium* spp.) and methanogens (*Methobactrium* spp.) that produce methane as a result of their metabolism. In contrast, eubacteria includes the commonly known bacteria (*E. coli*, *Staphylococcus aureus*, etc.) which commonly cause human infections, thus our focus here will be on the eubacteria.

### 1.1 Cellular Properties

Bacteria contain ribosomes in their cytoplasm, which consist of proteins and RNA and are needed for protein synthesis (Fig. 1). This function is similar between prokartotes and eukaryotes, however the actual components (i.e., RNA and proteins) are slightly different. These differences have proved very useful in combatting bacterial infections as many antibiotics are targetted against the protein synthesis function of the bacteria. The beauty is that although these antibiotics are highly effective in killing bacteria, they are useless against eukaryotic cells and thus do not exhibit any side effects. However, these antibiotics are obsolete against eukaryotic human pathogens. Bacteria contain DNA (a single chromosome) in the cytoplasm. The bacterial DNA vary in size between different species of bacteria (e.g., *E. coli* chromosome is approximately $5 \times 10^6$ bp). The DNA is circular and associated with proteins similar to histone proteins found in eukaryotic cells. Some bacteria also contain extra pieces of DNA in the cytoplasm (not part of the chromosomal DNA) which are called plasmids. The plasmid DNA is normally not crucial to the normal life of bacteria but may contain genes that produce proteins required for survival in harsh conditions, e.g., plasmids may carry genes for antibiotic resistance. The cytoplasm is enclosed within a plasma membrane (Fig. 2). It is a semi-permeable lipid-bilayer (around 5 – 10 nm thick) that acts as a barrier between cytoplasm and the external environment. It is made of phospholipids such as phospotidylcholine, which consists of a polar head group (hydrophilic) attached via glycerol to two long chains of non-polar fatty acids (hydrophobic) (Fig. 2). Such molecules consist of both polar and non-polar groups and are called amphipathic. The interior of the membrane is highly hydrophobic and does not allow the passage of anything that is bulky, even glucose. However, molecules such as water can pass through the hydrophobic interior. In addition, proteins are embedded in the lipid bilayer. The function of membrane proteins is to transport molecules across the membrane, energy metabolism and detect and respond to chemical stimuli and bind to specific host cells surfaces, etc. Proteins that are fully associated with the membrane (may penetrate all the way through) are

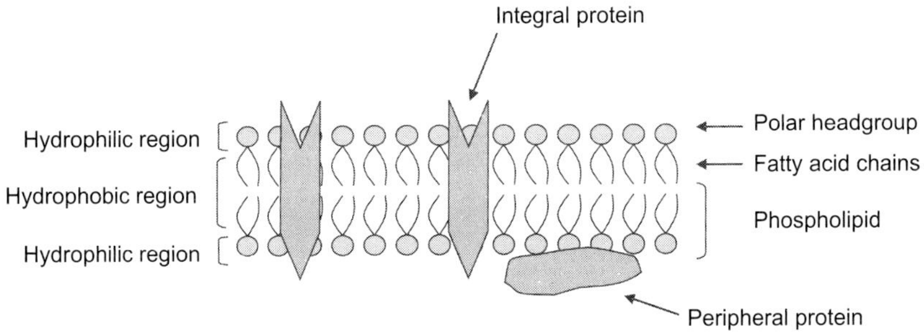


**Fig. 2** Structure of the cytoplasmic membrane.

called integral proteins (Fig. 2). These proteins therefore contain hydrophobic amino acids in the regions which are buried in the lipids. In contrast, proteins that are loosely associated by charge interactions are with the polar head groups of the phospholipids and are called peripheral proteins (Fig. 2). These can be removed from the membranes by washing with salts and detergents.

External to the plasma membrane is a rigid cell wall (Fig. 3). The cell wall gives shape to the bacterial cells but more importantly protects the bacteria from lysis, resulting from osmotic pressure, thus bacteria can survive in the environment or outside their host. For example, mycoplasmas do not possess cell walls and thus can not survive outside their host. The cell wall is made of sugars and amino acids. For example, the structure of *E. coli* consists of a long polymer of two sugar derivatives, N-acetylglucosamine (NAG) and N-acetylmuramic acid (NAM) with side chains of amino acids. Rigidity is achieved by cross-links between the amino acid chains. The unique feature of this membrane is the presence of NAM since this sugar is not found in the eukaryotic cells and provides a target for anti-bacterial therapy. For example, penicillin inhibits cross-linking and thus interferes with the peptidoglycan biosynthesis. In addition, Gram-positive bacteria contain a large amount of teichoic acid, made of glycerol and ribitol, and are attached to the NAM or lipids in the membrane (in which case it is called lipoteichoic acid). The cell walls of other genera such as *Mycobacterium* spp., *Cornebacterium* spp., *Nocardia* spp. possess waxy esters of mycolic acids, which are complex fatty acids. The cell walls of Gram-negative bacteria possess an additional layer called the outer membrane, which protects the peptidoglycan (Fig. 3). The space between the two membranes is called periplasmic space, which is a gel-like structure with a loose network of peptidoglycans. The periplasmic space contains various proteins involved in the transport of nutrients, enzymes that are important for nutrient acquisition, i.e., proteases, and enzymes that function against toxic chemicals, i.e., beta-lactamases which destroy the penicillin. The outer membrane of Gram-negative bacteria is made up of phospholipids and proteins that form pores allowing the passage of small molecules into

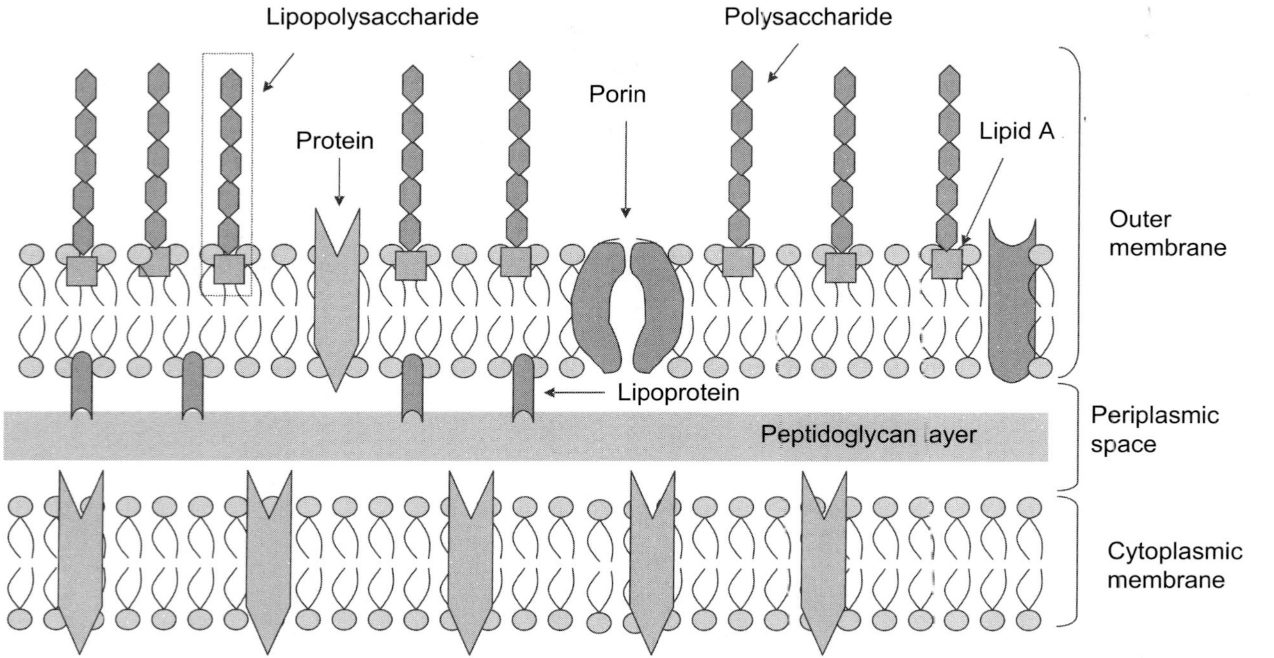


**Fig. 3** Membrane in Gram-negative bacteria. Gram-positive bacteria lack the outer membrane. The cell wall in Gram-negative bacteria is the outer membrane plus peptidoglycan, while in Gram-positive bacteria, it is a thick peptidoglycan layer only.

the periplasmic space and lipopolysachharide (see sections 1.3, 1.4, 3.1) on the outer leaflet of the outer membrane. External to the cell wall is a dense structure called capsule or glycocalyx. Capsules are made of polysaccharides but some capsules also possess proteins and are covalently linked to phospholipids or lipid A molecules. Capsules mediate a range of biological processes and may act as permeability barriers, protect bacteria from phagocytosis, prevention of desiccation, adherence and resistance to the host immunity. In addition, bacteria may possess movement appendages on their surface known as flagella, fimbriae or pili (Fig. 1). These structures aid in the bacterial movement as well as help bind to the susceptible surfaces. Some bacteria (e.g., *Bacillus* spp. and *Clostridium* spp.) can produce endospores, which contain a number of layers surrounding the bacterial genetic material. This enables the bacteria to become highly resistant to harsh conditions such as heat, UV, chemical disinfectants, drying and boiling. Among many bacteria, there are two major groups, Gram-positive bacteria and Gram-negative bacteria.

## 1.2 Gram-positive and Gram-negative Bacteria

The basis of this differentiation is a staining method devised by Christian Gram in 1884. Briefly, bacteria are fixed on a slide and stain with dark crystal violet and all cells become purple. Then, iodine is added which forms complex with crystal violet and enhances staining but all cells remain purple. Next, alcohol is added which removes the stain from the bacteria with porous walls, i.e., Gram-negative become colourless while Gram-positive retain their purple stain. Finally a paler stain such as safranin is added which stains the Gram-negative as pink (may appear pale purple), while Gram-positive cells remain dark purple under light microscopy. The basis of this method is that Gram-negative have a higher lipid content than Gram-positive bacteria. The increased lipid content in Gram-negative walls make them more porous and incapable of retaining the dye. The structural differences between Gram-positive and Gram-negative bacteria are shown in Fig. 1. As observed, the cell wall in Gram-positive bacteria is composed of the peptidoglycan layer, while Gram-negative bacteria possess an outer membrane in addition to the peptidoglycan layer.

## 1.3 Transport in Bacteria

Bacteria use several mechanisms to transport molecules in and out of the cell. If molecules move across the membrane without interacting with the proteins in the membrane, it is called passive transport (or passive diffusion) and is driven by the concentration gradient, e.g., $O_2$ and $CO_2$. However, the movement of large molecules require the function of membrane transport proteins (which usually span the entire membrane and are also known as transmembrane). This is similar to passive transport but is called facilitated transport.

The movement of molecules irrespective of their concentration in the environment is called active transport. This is an energy-dependent process and particularly useful for bacteria to survive in the environments with low nutrients. The substrate binds to a specific transport protein and during transport, the protein undergoes structural changes resulting in low affinity, and substrate is released into the interior of the cell (Fig. 4). The energy for this process is obtained from the proton motive force or ATP hydrolysis. In Gram-negative bacteria, the molecules are moved through pores into the periplasmic space. The binding proteins located in the peirplasmic space bind to the substrate. The protein-substrate complex then binds to proteins embedded in the inner membrane for active transport. Other forms of transport includes group translocation in which the substrate is chemically modified during transport, e.g., phosphorylation.

## 1.4 Bacterial Movement

Many bacteria use flagellum for their motility (Fig. 1). It is a long thin structure (3 to 12 μm long and 12 to 30 nm in diameter) which is free on one end and embedded in the cell on the other end. The position and

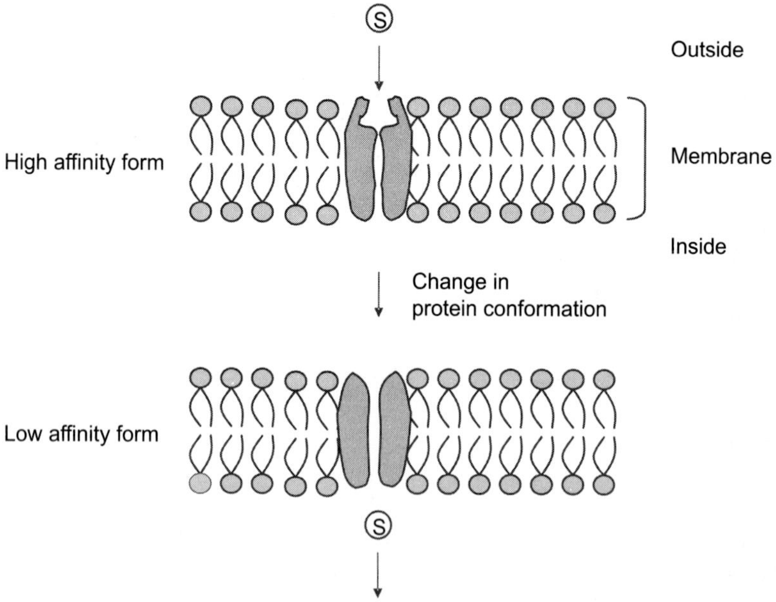


**Fig. 4** Active transport showing conformational changes in the transport protein.

number of flagella vary between different bacteria. Bacteria with a single flagellum are called monotrichous, several located on one end of the cell are known as lopotrichous, and several on both ends are termed amphitrichous and several all over the cell are known as peritricous (Fig. 5).

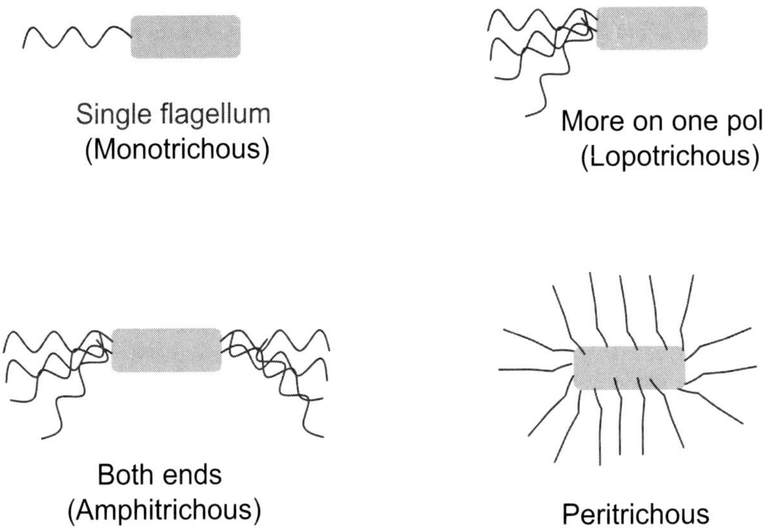


**Fig. 5** Bacterial types based on flagella.

Flagella are made of protein subunit called flagellin. Approximately 50 genes are required for the synthesis of flagellum. The part of flagellum that is embedded in the membrane possesses rings (Fig. 1 and Fig. 6). In Gram-negative bacteria, one pair of rings is associated with LPS and the peptidoglycan layer while other is associated with the inner membrane (Fig. 6). In Gram-positive bacteria that lack LPS, only one pair of rings is present. Flagella are rigid structures that do not flex but rotate, driven by the basal body together with rings acting as a motor. The rotation of flagellum determines the type of movement. For example, when a single flagellum rotates counter-clockwise, the bacteria moves forward and tumbles when it rotates clockwise. The bacteria uses this motion during chemotaxis. For example, when the bacteria sense food substrates (using their sensory proteins), they move towards the substrate using a counter-clockwise motion of their flagella, called positive chemotaxis. While moving away from the toxic molecules using a clockwise motion of their flagella. In addition, the bacteria may contain pili or fimbriae, hair-like appendages that are present on the surface of many Gram-negative cells. There are two types of pili, i) small number of pili (one to six) that are very long and known as sex pili that act as receptors for binding to bacteriophages, and

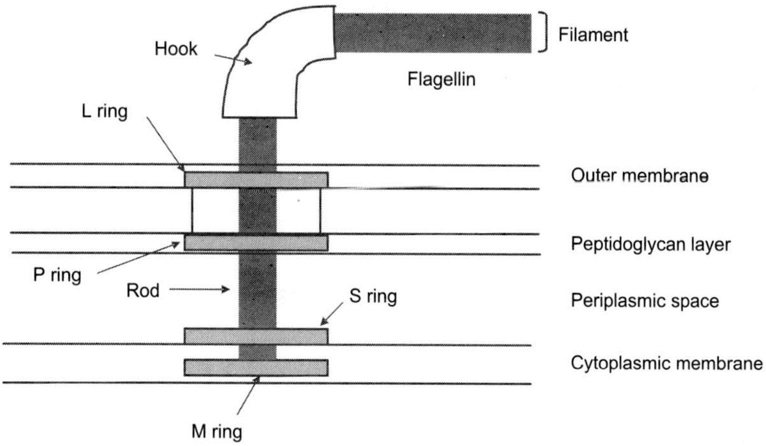


**Fig. 6** Structure of the flagellum basal body and hook in a Gram-negative bacterium.

ii) short, abundant common pili which are present in large numbers, e.g., *Shigell* spp. or *E. coli* that possess as many as 200 pili per cell. Pili also play a role in binding to the host cell and other surfaces.

## 1.5 Bacterial Growth

The bacteria divide by binary fission, in which the cyoplamic constituents are made and the genetic material is replicated and bacteria divides into two cells of equal size. The doubling times vary between different bacteria. For example, under favourable conditions *E. coli* can be divided into two cells within 20 min, while *Mycobacterium tuberculosis* takes around 18 h. The ability of the bacteria to grow rapidly is an important feature in colonizing and infection as well as developing resistance against antimicrobials (Table 1). The complete sequence of events involved in the cell division is called the cell cycle. It involves three stages, (i) C-phase (chromosome replication phase) during which the chromosome is replicated, (ii) G (gap)-phase (chromosome segregation phase) during which the replicated DNA binds to the two adjacent sites on the membrane (Fig. 7). As the membrane grows between the sites, it pushes the DNA towards the poles of the cells. Lastly, (iii) a cross wall (septum) is formed between the two chromosomes and the cell divides into two cells (D-phase) (Fig. 7). Both DNA replication and cell division is well coordinated. A cell must reach a critical mass before the initiation of DNA replication at the origin site (*oriC*), which is a short adenine and thymine rich sequence and continues bidirectionally round the chromosome to the termination point (Fig. 8). The phases in bacterial growth include a (i) lag-phase, when bacterial growth are newly inoculated into a medium and they are adapting

**Table 1.** *E. coli* growth under favourable conditions.

| *Time* | *No. of cells* |
|---|---|
| 0 | 1 |
| 20 | 2 |
| 40 | 4 |
| 1 hr | 8 |
| 1 hr 20 min | 16 |
| 1 hr 40 min | 32 |
| 2 hr | 64 |
| 2 hr 20 min | 128 |
| 2 hr 40 min | 256 |
| 3 hr | 512 |
| 4 hr | 4096 |
| 5 hr | 32768 |
| 6 hr | 262144 |
| 7 hr | 2097152 (2 million) |
| 8 hr | 16777216 (16 million) |
| 9 hr | 134217728 |
| 10 hr | 1073741824 |
| 12 hr | 68719476736 |

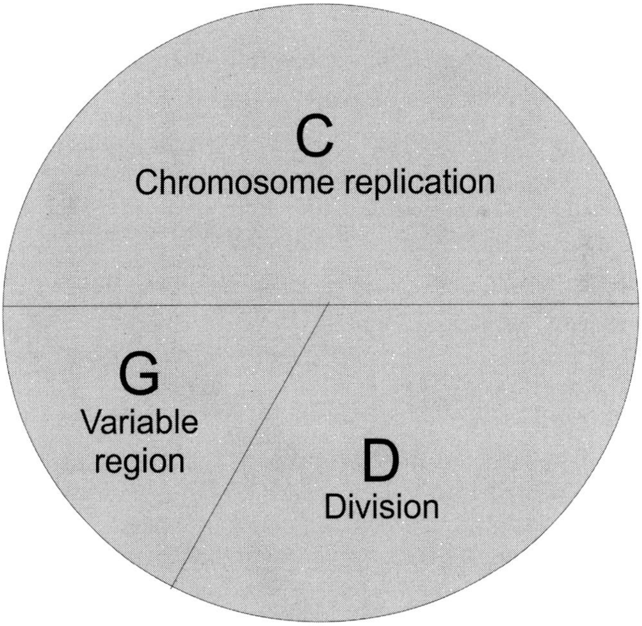


**Fig. 7** Stages in the cell cycle of prokaryotes.

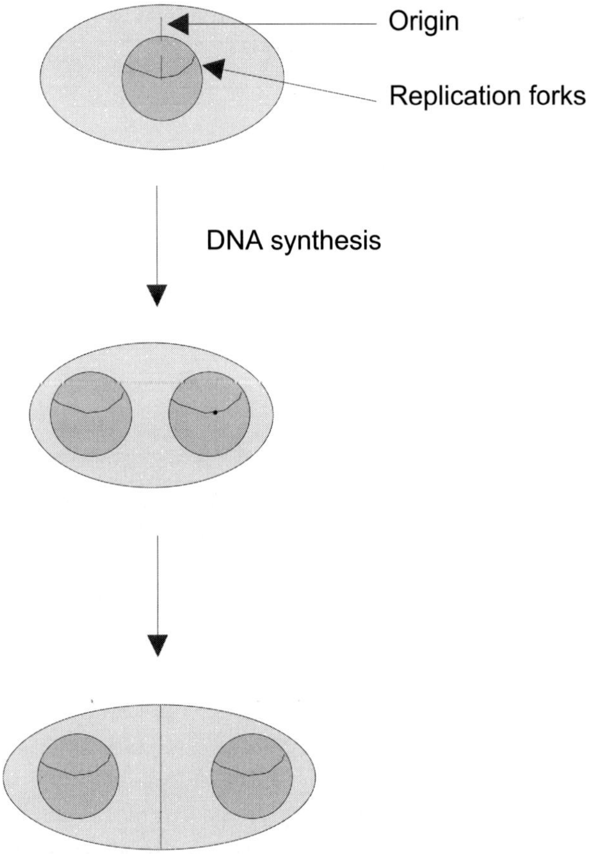


**Fig. 8** DNA replication and cell division.

to the new conditions before cell division, (ii) exponential (logarithmic-phase), when the bacteria start to divide and increase at a constant rate, (iii) stationary-phase, during which all the nutrients in the medium are used up and the number of bacteria reach a stationary point (Fig. 9). The number of bacteria which are killed equals the small numbers of bacteria that are being synthesized and finally (iv) the death-phase, during which rate of death becomes greater than cell division.

## 1.6 Bacterial Metabolism

Metabolism is the sum of total biochemical reactions required to maintain functional cellular activities. These reactions include degradative or breakdown processes resulting in small molecules and are called catabolic reactions, and the synthetic processes resulting in large molecules and are

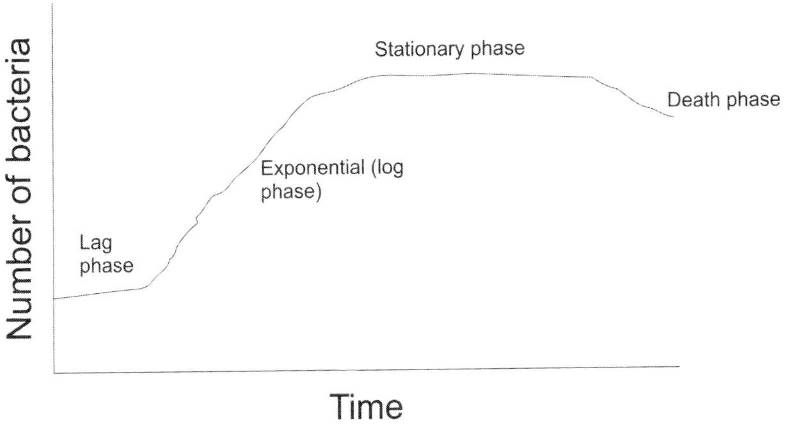


**Fig. 9** A typical bacterial growth curve.

called anabolic reactions. Both processes occur simultaneously and are the basis to provide nutritional needs of the cell. The bacteria can utilize a variety of food sources and uses diverse mechanisms to generate energy (Table 2). The energy generated by these reactions is usually stored by the formation of high energy compounds such as adenosine diphosphate (ADP) and adenosine triphosphate (ATP). Based on metabolic pathways, the bacteria can be divided into three groups.

**Table 2** Bacterial energy source.

| *Type* | *Carbon source* | *Nitrogen source* | *Energy source* |
|---|---|---|---|
| Heterotrophs | Organic | Organic or inorganic | Oxidation of organic compounds |
| Autrophs | $CO_2$ | Inorganic | Oxidation of inorganic compounds |
| Phototrophs | $CO_2$ | Inorganic | Sunlight |
| Cyanobacteria | $CO_2$ | Inorganic | Sunlight |

### 1.6.1 *Phototrophs (photosynthetic bacteria)*

It is a light-dependent process where carbon dioxide is reduced to glucose, which is then used for energy production.

### 1.6.2 *Autotrophs*

In this, the bacteria oxidize inorganic compounds (e.g., $NH_3$, $NO_2$, $S_2$, $Fe^{2+}$) without using sunlight to yield energy.

### 1.6.3 *Heterotrophs (also called chemo-organotrophs)*

These include nearly all pathogenic bacteria and are the focus here. These bacteria obtain energy by oxidizing organic compounds such as sugars (preferable food source), lipids and proteins (Fig. 10). To achieve this, bacteria secrete enzymes (glycosidases) that can degrade complex sugars into simple molecules, enzymes (phospholipases) that can degrade lipids and enzymes (proteases) which can degrade proteins. Under aerobic conditions, $O_2$ serves as the terminal acceptor of electrons (see below), while under anaerobic conditions, $NO_3^-$, $SO_4^{2-}$, $CO_2$, or fumarate can serve as a terminal electron acceptor. Overall, this results in the synthesis of ATP as the energy source. As an example, glucose oxidation is described below.

$$C_6H_{12}O_6 + 6O_2 \rightarrow 6CO_2 + 6H_2O + \text{Energy}$$

The net result of complete oxidation of a single glucose molecule is the production of 38 ATP molecules. If the substrate is protein, then ammonia is also produced.

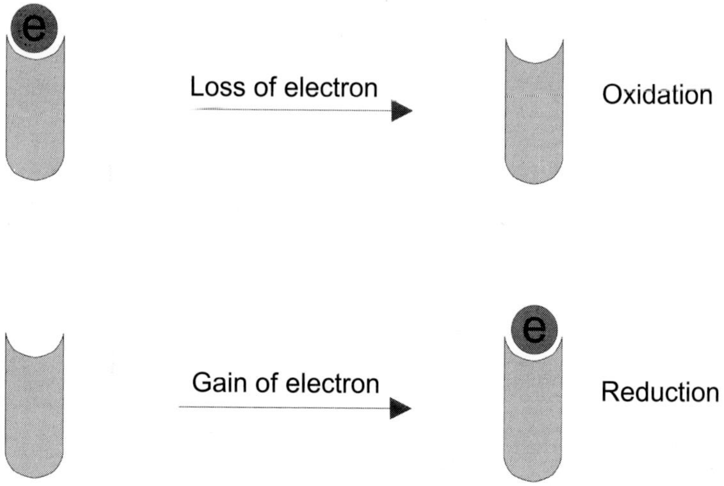


**Fig. 10** Oxidation and reduction reaction. Reduction describes the overall electrical charge. As the electrons are negatively charged and thus acceptors are referred to as reduced.

Glucose is catabolized via two processes, oxidation (complete breakdown of glucose into $CO_2$ and water) or fermentation (results in organic waste products) (Fig. 11). The entire oxidation process can be divided into the following three biochemical pathways including glycolysis pathway, the Krebs cycle and oxidative phosphorylation as described below.

**1.6.3i Glycolysis pathway (also called Embden-Meyerhof-Parnas pathway)** Glycolysis is used both by bacteria and eukaryotes to extract carbon and energy from glucose. Glucose is transported into the cell and phosphorylated by phosphoenolpyruvate (PEP) and transforms into glucose-6-phosphate by group translocation. Glucose can also be transported into the cytoplasm by active transport without being

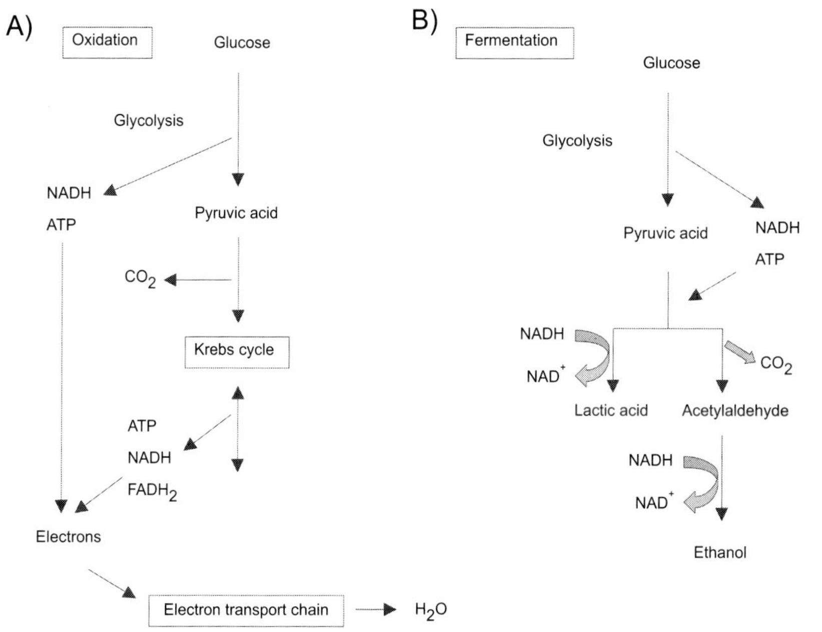


**Fig. 11** Summary of glucose metabolism indicating two pathways which branch from pyruvic acid.

phosphorylated. Glucose molecules transported via active transport are then phosphorylated by hexokinase enzymes, which transfer a phosphate group from an ATP molecule to the glucose resulting in 6C (carbon) glucose-6-phosphate. Glucose-6-phosphate is then phosphorylated again to 6C fructose-1,6-bihosphate (Fig. 12). Next, 6C fructose-1,6-biphosphate into two 3C pyruvate molecules (Fig. 12). In the process, two ATP are used while four ADP molecules are phosphorylated to make four ATP, i.e., net gain of 2ATP and two molecules of nicotinamide adenine dinucleotide ($NAD^+$) are reduced to NADH.

**1.6.3ii Krebs cycle (also called citric acid cycle or tricarboxylic acid cycle)** Before pyruvic acid can enter the Krebs cycle, it must be converted into acetyl-coenzyme A (acetyl-CoA) (Fig. 13). In brief, enzymes remove one carbon from the pyruvic acid as $CO_2$, and joining it with coenzyme-A resulting in acetyl-coenzyme-A (acetyl-CoA), a process called decarboxylation (Fig. 13). Overall, two molecules of pyruvic acid will result in two molecules of acetyl-CoA and two molecules of $CO_2$, and two molecules of NADH. Finally acetyl-CoA enters into the Krebs cycle (Fig. 14). It occurs in the cytoplasm in prokaryotes and in mitochondria in the eukaryotic cells. The overall outcome is that, for two molecules of acetyl-CoA, two ATP molecules are produced and six molecules the NADH and two molecules of $FADH_2$ are produced.

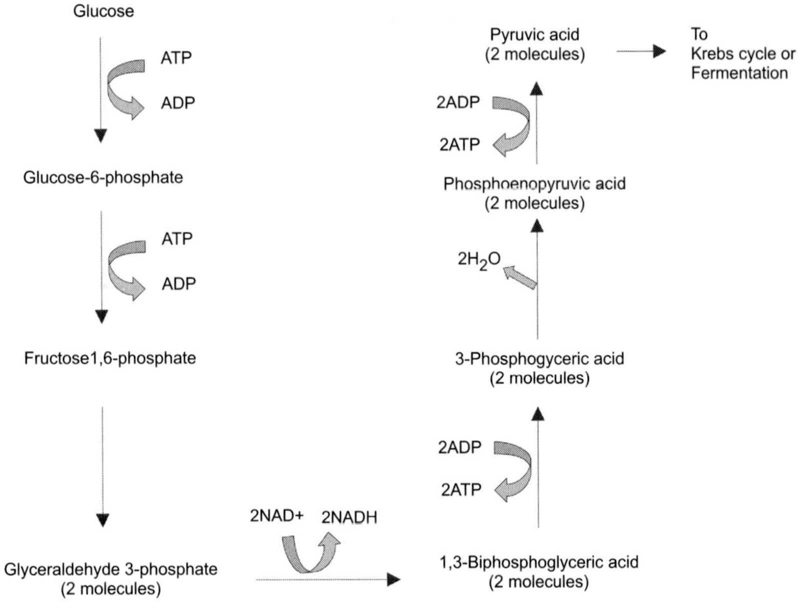


**Fig. 12** Glycolysis pathway.

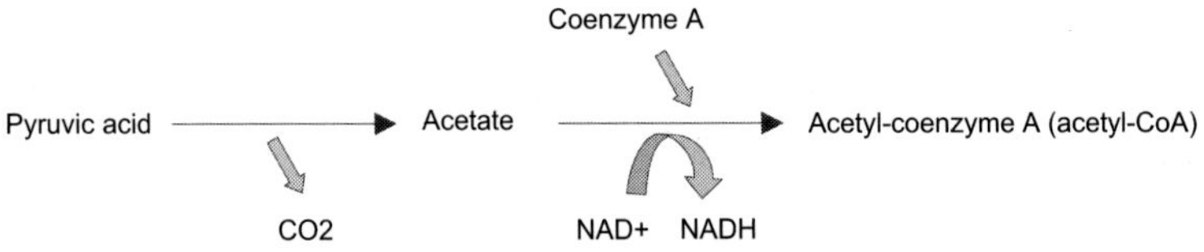


**Fig. 13** Formation of acetyl-coenzyme-A.

**1.6.3iii Electron transport** The electron transport system consists of a number of membrane-bound molecules (inner membrane of mitochondria in eukaryotes and cytoplasmic membrane in prokaryotes) that pass electrons from one another, ultimately to a final electron acceptor, while pumping protons across the membrane (Fig. 15). This process is associated with a protein complex, ATP synthase that generates ATP. The NADH and $FADH_2$ donate electrons, which are used to generate energy, which is captured to synthesize ATP (Fig. 15). The process initiates with NADH oxidation to $NAD^+$ transferring a proton and two electrons to a flavoprotein (flavin monnucleotide, FMN). Flavoproteins are integral-membrane proteins that alter between oxidized and reduced forms. Flavoprotein transfers the two electrons to the next protein, iron-sulphur protein (FeS) while releasing the two protons outside the cytoplasmic membrane. Iron-sulphur protein transfers electron to lipid ubiquinones which also picks up two protons as well, from the cytoplasmic side of the membrane. The ubiquinone carrying

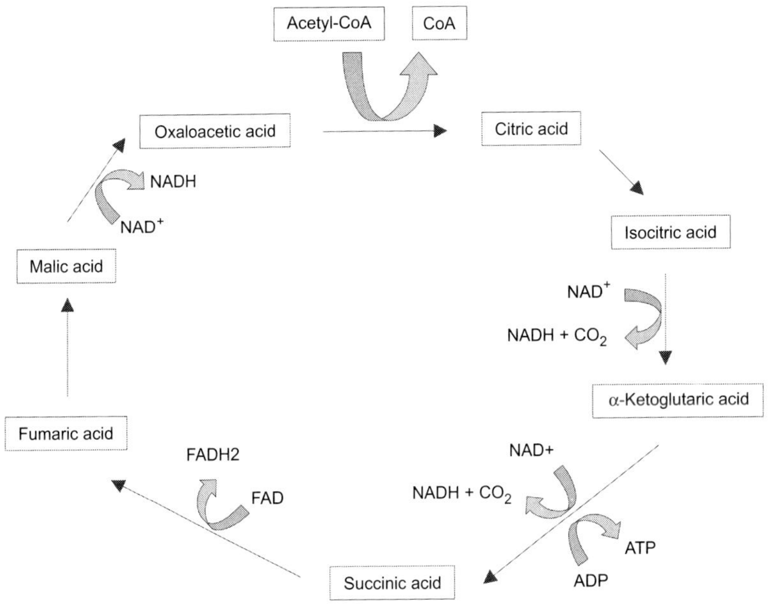


**Fig. 14** Krebs cycle.

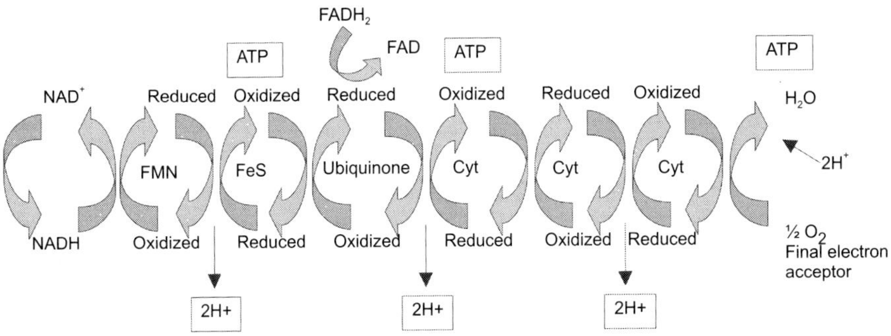


**Fig. 15** Electron transport chain.

protons and electrons interact with cytochrome and donate two electrons to cytochrome while releasing two protons outside the cytoplasmic membrane. Cytochrome oxidase finally oxidizes the cytochrome and transfers the electrons to the terminal electron acceptor, i.e., oxygen (nitrate, sulphate, or organic compounds can be electron acceptors under anaerobic conditions), which is converted to water by cytochrome oxidase. This process generates a proton gradient outside the cytoplasmic membrane, which is either used to rotate flagella or re-enter the cytoplasm through a complex called ATP synthase inducing the phosphorylation of ADP to

ATP. It takes around two protons to move through ATP synthase to drive phosphorylation of an ADP to ATP.

## 2. PATHOGENESIS AND VIRULENCE OF BACTERIAL INFECTIONS

There are hundreds of bacterial species causing hundreds of diverse bacterial infections. The basic pathogenic mechanisms associated with each bacterial infection involve diverse mechanisms and distinct virulence determinants. A complete description of each bacterial pathogen and their virulence determinants is beyond the scope of this book. Here, a common scheme relevant to the majority of bacterial infections is presented.

### 2.1 Adhesion

Adhesion is the primary step in bacterial pathogenesis. Certain bacteria may be limited in targetting specific tissues. For example, *Streptococcas mutans* is abundant in dental plaque but not on epithelial surfaces of the tongue. In contrast, *S. salivarius* binds to the epithelial cells of the tongue but is absent from dental plaque. This is due to the existence of precise interactions at the molecular level between specific bacterial determinants (known as adhesins) and the specific receptors (usually glycoproteins or glycolipids) on the host cells. For example, uropathogenic *E. coli* binds to the urinary tract epithelial cells using protein PapG on their fimbriae (also see section E). The ability of bacteria to bind to the host cells is crucial to remain in the host body. Nearly all surfaces in the body are protected by some form of flushing movements (e.g., intestinal tract and the bloodstream), which do not allow the bacteria to remain unattached and constantly attempt to flush them out. Most pathogenic bacteria express adhesins on their surface to allow the attachment of the bacterium to the host cell surfaces. Many adhesins are proteins and are expressed on the surface of pili/fimbriae (Fig. 1). The structures of the fimbriae are thought to help overcome the electrostatic repulsions resulting from the negative charge on both cells, i.e., bacterium and the host cell (Fig. 16). The receptor(s) may be specific glycoprotein or glycolipid which may be limited to specific cell types, thus providing a species, tissue or cellular tropism. Thus, bacteria that infect various tissues or different organisms may require different adhesins.

### 2.2 Colonization

A single bacterium is unlikely to produce an infection. The ability of bacteria to remain at a particular site and multiply is called colonization. It is

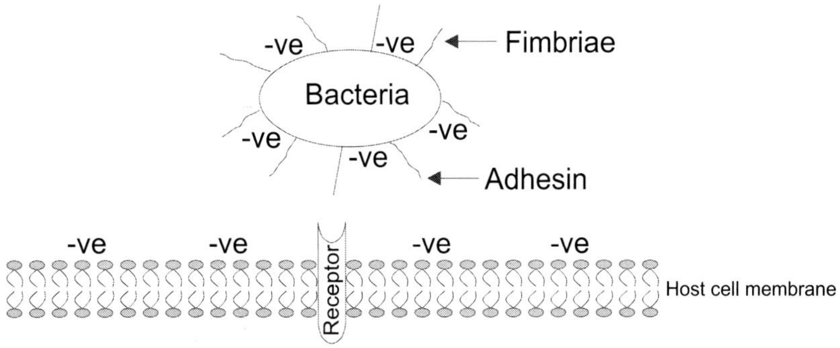


**Fig. 16** The role of pili/fimbriae in binding to host cell membranes.

essential for bacteria to grow in large numbers to persist, despite a competent host immune response. For some bacteria, colonization is sufficient to produce infection. For example, enterotoxigenic *E. coli* (ETEC) bind to intestinal epithelial cells and colonize. This is sufficient to produce toxins and cause infection, i.e., secretory diarrhoea. Similarly, enteropathogenic *E. coli* (EPEC) and enterohemorrhagic *E. coli* (EHEC) bind to the intestinal epithelial cells, colonize and secrete toxins to produce malabsorptive diarrhoea and dysentery (bloody diarrhoea). Thus binding and colonization is sufficient to produce infections in some bacteria.

## 2.3 Secretion of Toxins

Pathogenic bacteria produce toxins, which cause damage to the host cells. For example, following binding, ETEC produce toxins (heat-labile toxin, LT and heat-stable toxin, ST), which enter the intestinal epithelial cells and disrupt water and ion flow thus producing secretory diarrhoea. In contrast, EPEC bind to the intestinal epithelial cells and inject toxins into the host cytoplasm (directly from the bacterial cytoplasm into the host cytoplasm) producing malabsorptive diarrhoea. While, EHEC (O157:H7) inject toxins into the host cells to produce dysentery. In addition, EHEC produces notorious Shiga-like toxin, which enter the bloodstream and cause endothelial cell damage as well as kidney failure.

## 2.4 Entry into the Host Cells

The ability of some pathogenic bacteria to produce infection is not mere binding and colonization but they exhibit complex pathogenic mechanisms to produce the disease. This is particularly true for bacterial pathogens that produce central nervous system (CNS) infections. For example, the majority of meningitis-causing bacteria cross the gut, evade the immune responses, traverse the blood-brain barrier to gain entry into the CNS and produce the disease. To achieve this, the majority of the meningitis-causing bacteria have evolved mechanisms to invade the host cells. This ability allows bacteria to cross a biological barrier as well as protects them from an overwhelming immune response and to hide from antimicrobials circulating into the bloodstream. Again, different bacteria use diverse proteins/toxins to invade into the host cells. The bacterial invasion of the host cells involve host cell cytoskeletal rearrangements (Fig. 17). Cytoskeleton is a network of filaments just beneath the plasma membrane that provides structural support to the cells. The ability of the bacteria to manipulate cytoskeletal proteins to gain entry into the host cells is crucial to remain viable and cross the barrier for the onset of the disease.

## 2.5 Evasion of Host Killing Mechanisms and Intracellular Multiplication

Once taken up, the bacteria use diverse strategies to avoid host cell killing (Fig. 17). For example, *Listeria monocytogenes* invade the intestinal epithelial cells escape phagosome, multiply and move from one cell to another adjacent cell eventually reaching the bloodstream. In contrast, *Salmonella*

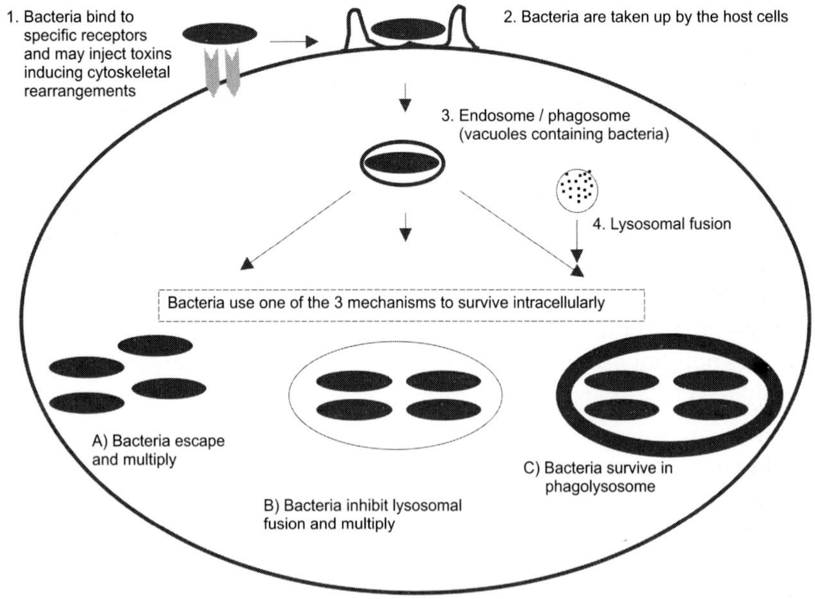


**Fig. 17** Bacteria induces cytoskeletal rearrangements resulting in its uptake. Once inside the host cell, the bacteria use diverse mechanisms (indicated by A, B, C) to evade intracellular killing.

spp. invade intestinal epithelial cells and then infect the underlying macrophages. Once inside the macrophages, the bacteria inhibit fusion of lysosomes with phagosome. Similarly, *Mycobacterium tuberculosis* inhibit fusion of lysosomes with phagosome in macrophages. Other bacteria such as *Staphylococcus aureus* produce catalase and superoxide dismutase which detoxify toxic oxygen radicals and survive macrophage onslaught (Fig. 17). Other mechanisms of avoiding host cell killing are to destroy the host and/or phagocytic cells. For example, *Streptococcus pyogenes* kill the phagocyte by secreting streptolysin that induces lysosomal discharge into the cell cytoplasm.

### 2.6 Evasion of the Host Immune Response

A key component in bacterial infections is their ability to evade the host immune responses. This property of bacteria is covered in detail in later sections.

## 3. BACTERIAL TOXINS

Traditionally, bacterial toxins are divided into two broad categories, endotoxins and exotoxins.

### 3.1 Endotoxins

These are cell-associated toxins that are structural components of the outer membrane of Gram-negative bacteria. For example endotoxins are part of the outer membrane of Gram-negative bacteria such as *E. coli*, *Salmonella*, *Shigella*, *Pseudomonas*, *Neisseria* and *Haemophilus*. In *E. coli*, endotoxin (lipopolysaccharide, LPS) is released from the cell in the bloodstream after lysis as a result of the action of the host immune system or by the action of some antibiotics. The LPS is a heat-stable toxin and is composed of three components (Fig. 18):

(i) O-polysaccharide, which is associated with immunogenicity of the bacteria, attached to the core polysaccharide but much longer than the core polysaccharide. It maintains the hydrophilic domain of the LPS molecule and is a major antigenic determinant of the Gram-negative bacteria (there are >170 different O-antigens).

(ii) Core polysaccharide – with some variations, the core antigen is common in Gram-negative bacteria.

(iii) Lipid A, which is associated with the toxicity of LPS. Lipid A is hydrophobic, membrane-anchoring region of LPS and its structure is highly conserved among Gram-negative bacteria.

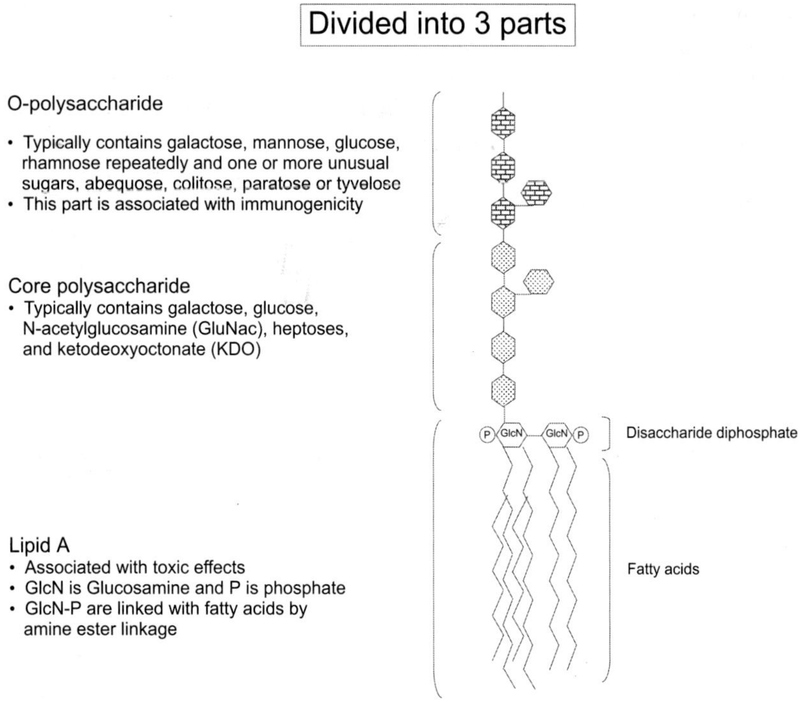


**Fig. 18** Structure of the endotoxin, lipopolysaccharide (LPS).

Complete LPS is called smooth or S-LPS while LPS lacking (or modified) O-antigen is known as rough LPS. O-antigens are targets for the action of the host complement system, but when the reaction takes place at the tips of the polysaccharide chains, complement fails to have its normal lytic effect on the bacteria. Some bacteria are virulent because of this resistance to immune forces of the host. If the projecting polysaccharide chains are shortened or removed, antibody reacts with antigens on the general bacterial surface, or very close to it, and the complement can lyse the bacteria. Hence the bacterial strains expressing 'rough' LPS are killed and are considered non-pathogenic.

### *3.1.1 Mechanisms of LPS action*

Once released in the serum, LPS binds to a serum protein called LPS-binding protein (LBP). This LPS-LBP complex binds to the CD14 receptor, which is primarily expressed on the surface of phagocytes (macrophages and neutrophils) and endothelial cells. These interactions induce the activation of a toll-like receptor-4 (TLR-4), leading to the stimulation of the intracellular signalling pathways thus stimulating the inflammatory response i.e., the release of cytokines such as interferon-gamma (IFN-$\gamma$), tumour necrosis factor-alpha (TNF-α) and interleukin-1 (IL-1) as well as oxidative bursts (Fig. 19). The low levels of inflammatory mediators have beneficial effects such as moderate fever and stimulate the host immune system, which lead to bacterial killing. However, overproduction of inflammatory mediators lead to the pathophysiological events such as high fever, hypotension, disseminated intravascular coagulation, increase in endothelial permeability leading to shock and could result in the host death. For example, the injection of purified LPS, into the experimental animals causes pathophysiological reactions such as fever, changes in white blood cell counts, disseminated intravascular coagulation, hypotension, shock and death. Although the Gram-positive bacteria lack endotoxin, they possess other components, i.e., lipoteichoic acids and peptidoglycans that can also induce septic shock and death.

## 3.2 Exotoxins

Exotoxins are proteins produced by both Gram-negative and Gram-positive bacteria and are toxic to the host cells. They are highly immunogenic and are usually secreted by the bacterial cells but may be held on the surface of the bacterial cell. Exotoxins are generally heat-labile (heat-sensitive) but there are some heat-stable toxins such as *E. coli* ST enterotoxin. In some

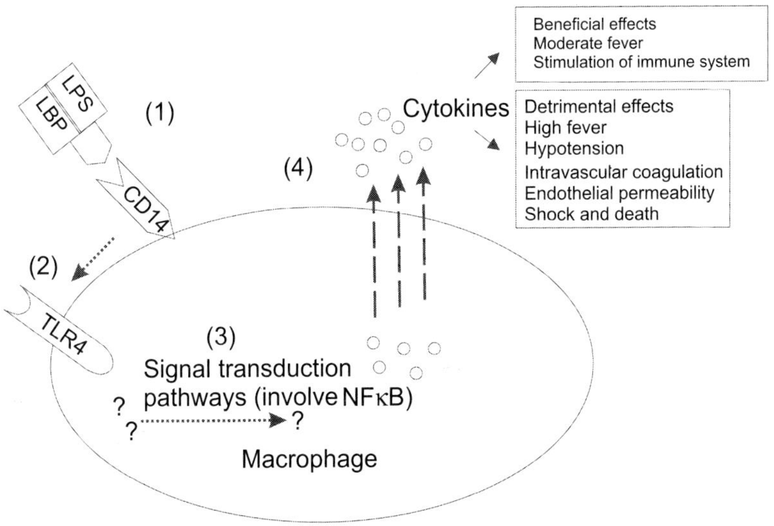


**Fig. 19** Mechanisms of LPS action and their role in disease. (1) LPS is released in the serum and binds to a serum protein, LPS-binding protein (LBP), which binds to CD14 on phagocytes. (2) CD14 activates toll-like receptor-4, which (3) leads to signal transduction pathways inducing (4) cytokine release.

cases (such as *E. coli* strains causing intestinal infections), these toxins are sufficient to produce infection. Some toxins have broad specificities such as phospholipases that cleave phospholipids which are regular components of the host cell membranes, resulting in the leakage of the cytoplasmic contents and cell death. While other toxins are highly specific in their actions. For example, *Clostridium botulinum* produces toxin that travels in the bloodstream to target neurones and produces botulism (flaccid paralysis). Similarly, the toxin of *C. tetani* produces tetanus, i.e., spastic paralysis. In contrast, toxin of *Corynebacterium diphtheriae* targets a variety of cell types and organs such as the heart and causes its failure, i.e., diphtheria.

## 3.3 Membrane-damaging Toxins

These toxins cause damage by destroying the integrity of the plasma membrane of the host cells. Usually they are active against a wide range of cell types and can be divided into two groups on the basis of their mechanisms of action which are described below.

### *3.3.1 Toxins with enzymatic action*

A large number of Gram-positive bacteria produce enzymes such as phospholipases and sphingomyelinases, which break down the lipid component on the plasma membrane. This destabilizes the membrane, eventually leading to the cell lysis.

### *3.3.2 Pore-forming toxins*

This group of toxins includes the thiol-activated cytolysins, streptolysin O (produced by *Streptococcus pyogenes*) and listeriolysin O (produced by *Listeria monocytogenes*). These toxins form pores in the host cell membranes. This results in the leakage of nutrients and essential ions from the cell and eventually results in the cell lysis.

## 3.4 Intracellular Acting Toxins

This is a potent group of toxins and act on the host intracellular targets. These include proteins such as botulinum toxin which inhibits the release of neurotransmitters in the peripheral nervous system, diphtheria toxin that inhibits protein synthesis resulting in cell death and the cholera toxin that acts in the gut to induce diarrhoea (Table 3). Intracellular acting toxins can be divided into three broad categories.

### *3.4.1 A-B toxins*

These are common types of toxins. One part of the toxin, B portion is responsible for binding to the specific receptors, while the A portion carries

**Table 3** Examples of bacterial exotoxins that act inside the cell.

| *Bacteria* | *Toxin* | *Site of action* | *Mode of action* | *Symptoms/role in disease* |
|---|---|---|---|---|
| *E. coli*<br>*Vibrio cholerae* | LT toxin<br>Cholera toxin | Intestinal epithelial cells | ADP-ribosylation of G protein which regulate adenylate cyclase | Water secretion into the intestine resulting in watery diarrhoea |
| *Corynebacterium diphtheriae*<br>*Psudomonas aeruginosa* | Diphtheria toxin<br>Exotoxin A | Many cell types | ADP-ribosylation of EF-2 leads to inhibition of protein synthesis | Cell death, general organ damage |
| *Bordetella pertussis* | Pertussis toxin | Many cell types | ADP-ribosylation of G protein which regulate adenylate cyclase | Cell damage, fluid secretion |
| *Shigella dysenteriae*<br>*E. coli* | Shiga toxin<br>Vero (Shiga-like) toxin | Many cell types | RNA glycosidase enzyme modifies 28S rRNA in 60S ribosome subunit. Inhibits protein synthesis | Cell death |
| *Clostridium tetani* | Tetanus toxin | Neurones in CNS | Metallopeptidase inhibits release of neurotransmitters | Spastic paralysis |
| *Clostridium botulinum* | Botulinum toxin | Peripheral neurons | Metallopeptidase inhibits release of neurotransmitters | Flaccid paralysis |

the toxic enzymatic activity. The diptheria toxin is one of the simplest of the A-B toxin consisting of just one A and one B portion. These are synthesized together as a single molecule, and then cleaved to give two peptides joined by a disulfide bridge. In contrast, the cholera toxin consists of five B subunits surrounding an active $A_1$-$A_2$ molecule, which is synthesized separately. The B part is non-toxic and is responsible for cellular specificity. It will bind only to cells that have the correct receptor. Hence in the case of botulinum toxin, the receptor is only present on neurones. After binding to a specific receptor, the toxin-receptor complex is internalized either by endocytosis, followed by translocation of the A portion from the endocytic vacuole into the cytoplasm or by translocation of the A portion across the host plasma membrane directly into the cytoplasm. In many cases, the enzymatic activity of A-B toxin is ADP-ribosylation of a target site in the cell. The A portion catalyzes the removal of the ADP ribosyl group from NAD and the attachment of that group to a cellular protein. The target proteins and the consequences of the ADP-ribosylation are variable thus explaining various types of symptoms. The diphtheria toxin, for example, inhibits protein synthesis, with one molecule enough to kill the cell. In contrast, the cholera toxin modifies G protein, a regulatory protein of adenylate cyclase causing it to be permanently switched to the production of cAMP. In the intestinal tract, where the toxin acts, the most significant effect of rise in cAMP is the production of an ion imbalance, leading to massive water loss from the cell, i.e., diarrhoea.

### *3.4.2 Superantigens*

These toxins attach to the T cell receptors of T helper cells and the major histocompatibility complexes (MHC) of the antigen presenting cells (macrophages). This results in the hyper-activation of T helper cells (more than normal) without their recognition of an antigen (Fig. 20). The activation of so many T cells results in the release of large amounts of cytokines which can cause a symptom similar to septic shock. For example, *Streptococcus pyogenes* produce superantigens which produce extensive tissue damage around a wound (cellulitis).

### *3.4.3 Toxins secreted by type III secretion system*

This is a recently discovered protein (toxin) secretion system in Gram-negative bacteria. It enables Gram-negative bacteria to inject proteins

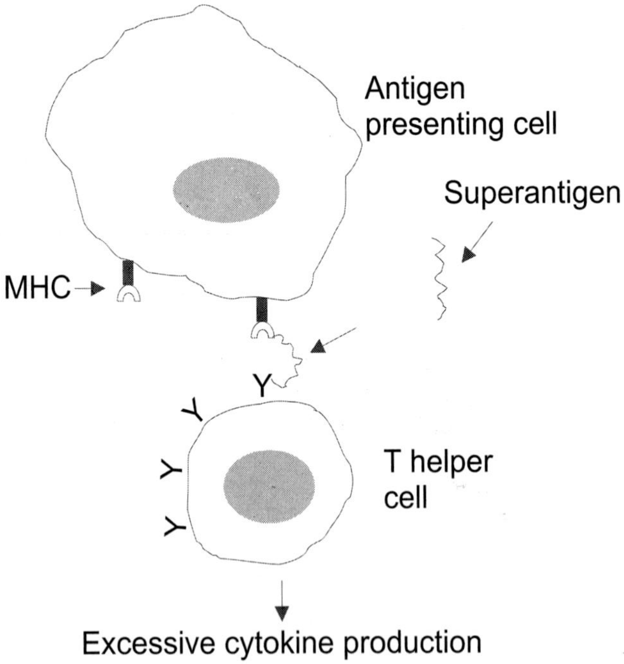


**Fig. 20** Superantigens mediate MHC (of antigen presenting cells but without an antigen) binding to T helper cell. This leads to hyper-activation of T cells resulting in excessive the cytokine release, which may result in shock.

directly from their cytoplasm into the host cell cytoplasm. Once the bacteria make a close contact with the host cell, it forms a tube between itself and the host cell. Many different proteins are required to build this organelle and consist of a thin tube with collars, which is inserted into the bacterial cell membranes and the eukaryotic host cell membrane. Following this, toxins are directly injected from the bacterial cell cytoplasm into the host cell cytoplasm. These toxins (proteins) are not subjected to N-terminal processing during secretion, thus functions differently from a common secretory pathway. In addition, these toxins do not require a 20 amino acid signal sequence required for secretion but in most cases this process seems to be regulated by bacterial contact with the target cell. The injected proteins interfere with the host intracellular signalling pathways resulting in distinct phenotypes. For example, when *Yersinia* encounters a macrophage, they inject toxins (YopH) into the host cell which interfere with signalling pathways and inhibit bacterial uptake. In contrast, *Salmonella* uses type III secretion system to inject toxins (SopE) in the intestinal epithelial cells which produces cytoskeletal changes and induce bacterial uptake by the host cells.

## 4. BACTERIAL EVASION OF IMMUNE DEFENCES

As indicated above, humans have developed a highly professional immune system to defend against bacterial or other microbial pathogens. As bacteria evolve faster than humans they have developed a variety of strategies to overcome and/or evade these defence systems. An overview of bacterial strategies to evade human immune responses is provided below.

### 4.1 Evasion of Antibodies

At the mucosal surfaces, bacteria have to cope with potent secretory immunoglobulins (sIgA) and then again in the bloodstream, IgG. These and other antibodies are highly effective in trapping microbes and neutralizing toxins. The bacteria exhibit the following evasion mechanisms to inactivate antibodies (Fig. 21).

#### *4.1.1 Glycosidases*

Glycosidases are enzymes that cleave carbohydrate chains. Antibodies such as sIgA are heavily glycosylated and glycosylation is essential for antibody function, conformation and net charge. Some pathogenic bacteria secrete glycosidases removing carbohydrate chains from antibodies rendering to their inactivity.

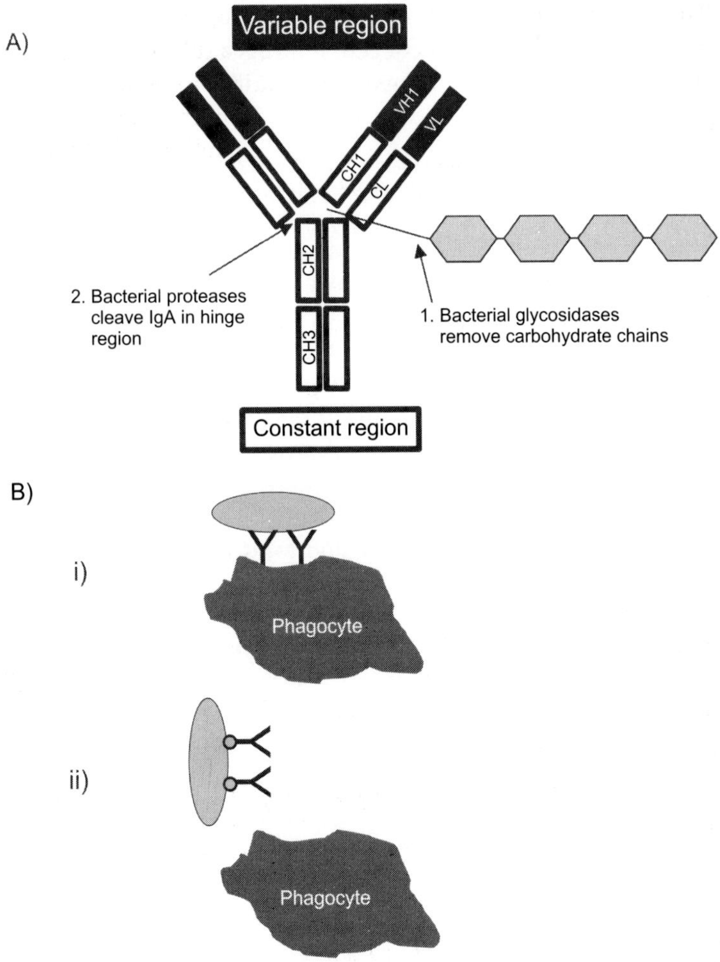


**Fig. 21** Bacterial evasion of antibodies. A) Bacterial glycosidases cleave carbohydrate chains rendering them obsolete. B) Bacterial proteases cleave antibody in the hinge region. C) Phagocytes normally bind to the Fc domain of the antibody but some bacteria produce Fc-binding proteins thus preventing phagocyte binding.

### *4.1.2 Proteases*

Some pathogenic bacteria produce proteases (called Ig proteases) which cleave Ig in the hinge regions – between CH1 and CH2.

### *4.1.3 Antibody-binding proteins*

Some pathogenic bacteria produce Ig binding proteins. Such proteins bind to the Fc region of Ig thus inhibiting phagocytosis (Fc regions are required for binding to the phagocytes). For example, *Staphylococcus aureus* produces cell wall-anchored protein, Protein A, which binds to the Fc domain of the

IgG. Thus, the antibody cannot react with Fc receptors on phagocytes and bacteria gets away. Protein G and Protein H are additional Streptococcal IgG binding proteins that result in the immune evasion.

#### 4.1.4 Antigenic variation

Some pathogenic bacteria shed old antigens and present new ones to the immune system making the available immune response obsolete. New antigens (fimbriae or other outer membrane proteins) do not bind to the existing antibodies. Thus, the host must produce new antibodies to challenge new antigens. The antibody production may take several days providing bacteria sufficient time to colonize and produce infection. Antigenic variations usually results from site-specific inversions or gene rearrangements in the DNA of microorganisms. *Neisseria gonorrhoeae* can change fimbrial antigens during the course of an infection.

## 4.2 Evasion of Cytokines

Cytokines are smoke signals that warn the human body of the invading bacteria, help attract phagoctes to the site of infection as well as link non-specific immune response with the specific immune response and thus help build an effective defence system against the invading bacteria. There are several strategies employed by the bacteria to interfere with cytokine synthesis, release and function as indicated below.

#### 4.2.1 Bacterial toxins inhibiting cytokine synthesis

Lymphostatin protein produced by enteropathogenic strains of *E. coli* inhibits the production of interleukin-2 (IL-2), IL-4 and interferon-gamma (IFN-$\gamma$). *Yersinia* can inhibit tumour necrosis factor-alpha (TNF-$\alpha$) synthesis by injecting toxins (Yop proteins).

#### 4.2.2 Binding to cytokines

*Escherichia coli*, *Shigella* spp. and *Mycobacterium tuberculosis* have receptors for cytokines such as IL-1, (TNF-$\alpha$) and epidermal growth factor (EGF) thus interfering with their function.

#### 4.2.3 Inactivating cytokines

Bacterial proteases can directly inactivate cytokines or cleave cytokine receptors on host cells so they can not respond or communicate with other immune cells to build an effective immune response.

#### 4.2.4 Induction of anti-inflammatory cytokines response

Some bacteria can induce an anti-inflammatory response by induction of IL-10 and transforming growth factor-beta (TGF-$\beta$) production, which have immunosuppressive effects as well as inhibit macrophage activation.

## 4.3 Evasion of Complement

The compement is a set of proteins that circulate in the blood and are activated through antibody binding to the bacterial surface or directly by bacterial surface leading to a MAC attack (forming pores in the bacterial cell membranes) or opsonises bacteria (help bacterial uptake by the phagocytes).

### 4.3.1 *Inactivating complement proteins*

*Streptococcus agalactiae* encode enzymes responsible for inactivating C5a. *Pseudomonas aeruginosa* produce elastase that inactivates C3b and C5a.

### 4.3.2 *Avoiding complement by binding to protectins*

To avoid host cells from unnecessary complement lysis, the host produces proteins, called protectins, which bind to the complement proteins and do not allow this process to proceed. They are also required to keep the complement system in check. It is shown that *Streoptococcus pyogenes* can bind to protectins and can avoid complement lysis.

## 4.4 Evasion of Phagocytic Killing

As indicated in section F, phagocytes (neutrophils and macrophages) play a crucial role in uptaking, killing and clearing of the bacteria from the human body. However, the pathogenic bacteria employ a diverse array of mechanisms to evade the uptake and onslaught by the phagocytes as described below.

### 4.4.1 *Avoiding contact with phagocytes*

Many pathogenic bacteria invade or remain in regions inaccessible or less patrolled by phagocytes, e.g., internal tissues such as the CNS and surface tissues (e.g., skin) are less guarded by phagocytes.

### 4.4.2 *Avoid provoking an overwhelming inflammatory response*

Some bacteria inhibit phagocyte chemotaxis, *Mycobacterium tuberculosis* inhibit leukocyte migration. *Clostridium* toxin inhibits neutrophil chemotaxis.

### 4.4.3 *Hide antigenic surface of bacterial cell*

Bacteria cover their surface with components seen as 'self' by host phagocytes and other immune systems. For example, pathogenic *Staphylococcus aureus* produces cell-bound coagulase which clots fibrin on the bacterial surface.

### 4.4.4 *Inhibition of phagocytic engulfment*

Many pathogenic bacteria bear molecules on their surfaces which inhibit phagocytosis. For example, polysaccharide capsules of *S. pneumoniae, Haemophilus influenzae, Treponema pallidum* and *Klebsiella pneumoniae* may inhibit phagocytic engulfment.

### 4.4.5 *Escape from the phagosome*

If taken up, bacteria are contained in a vacuole, i.e., phagosome. Early escape from the phagosome vacuole is used by some bacteria for their growth and virulence. For example, *Rickettsia* spp. produce phospholipase which lyse phagosomal membranes within 30 seconds of ingestion and thus evade phagocytic killing.

### 4.4.6 *Inhibition of fusion of lysosomes with phagosome*

Some pathogenic bacteria can even survive inside phagosomes because they prevent discharge of lysosomal contents into the phagosomal environment. For example, phago-lysosome formation is inhibited in the phagocyte by *Salmonella* and *M. tuberculosis*. With *M. tuberculosis*, the bacterial cell wall components are released from the phagosome and modify lysosomal membranes to inhibit lysosomal fusion. Many pathogenic bacteria produce enzymes i.e., superoxide dismutase that convert superoxide to water and thus survive the macrophage myeloperoxidase system. *Coxiella burnetii* is an intracellular bacterium that is phagocytosed by the macrophages, followed by the fusion of lysosomes to form phago-lysosome (low pH). The precise mechanisms are unclear but bacterium can survive and grow at acidic pH.

### 4.4.7 *Killing phagocytes before ingestion*

Other strategies of evasion of phagocytes are to attack and kill the phagocytes before they can attack bacteria. Some bacteria secrete extracellular enzymes that kill the phagocytes.

### 4.4.8 *Killing phagocytes after ingestion*

Some pathogenic bacteria may grow in the phagosome and release substances which can pass through the phagosomal membrane and cause discharge of lysosomal granules into the host cell, thus producing damage to the host cell.

## 5. CONTROL OF BACTERIAL INFECTIONS

Despite our efforts to control and/or eradicate pathogenic bacteria, they have continued to cause millions of deaths every year and remain a major contributing factor to human misery with both economical and social implications. The importance of sanitation, good housing, nutrition and clean water supplies is paramount in reducing the number of bacterial infections. For example, even before the introduction of vaccines and antibiotics, the reduction in the number of infections due to tuberculosis was noticed with improved public health measures. Other than improved public health measures, vaccines and antibiotics are the major weapons in our fight against bacterial diseases.

## 5.1 Vaccines

Vaccination is one of the most powerful means to save lives and to increase the level of health in mankind. Vaccination has proven a useful tool for protecting individuals and the population from bacterial diseases. A vaccine is a material originating from a microorganism (e.g., bacteria, attenuated bacterium, killed bacterium, denatured toxin, capsule, etc.) that is introduced into the individuals in a controlled way leading to the stimulation of the immune response without the symptoms of a full-blown disease. Material may be produced artificially or directly obtained from the pathogen. Ultimately, this results in the production of memory cells within the host. If the host encounters the pathogen again, the immune system can generate a rapid antibody response thus preventing infection. Vaccines are produced with the following objectives:

1. A vaccine should promote an effective resistance to disease.
2. It should provide sustained protection – lasts several years.
3. Induces humoral response (antibodies).
4. Induces protective T-cell responses (cellular immunity).
5. A vaccine must be safe – minimal side effects.
6. A vaccine should be stable and should remain so during transportation.
7. A vaccine should be reasonably cheap and easily administered.

However, not all vaccines fit these criteria and trials have to be conducted to determine whether the beneficial effects outweigh the potential side effects. The following types of vaccines are available:

### *5.1.1 Whole cell vaccines*

- Killed pathogen (ruptured or formalin-preserved), these vaccines should stimulate maximum protective immunity but may have severe side effects.
- Live attenuated or low virulence vaccines, these can be produced by knocking-out the virulent genes from the pathogen but may have side effects.

### *5.1.2 Protein subunit vaccines (single /multiple components)*

Purified proteins from the pathogen; these may induce maximum protection but may be toxic. Also, protein purification is expensive. In contrast, recombinant protein antigens are cheap and consistent. Other problems may arise from the fact that pathogen populations might be polymorphic. A vaccine may be ineffective against some strains. A simple recombinant might not stimulate all required immune components, e.g., immunogenic carbohydrate wrongly processed in production.

### *5.1.3 DNA vaccines (nucleic acid vaccines)*

For these, DNA (gene) encoding a multiple vaccine protein is cloned into a plasmid containing a host expression promoter. Individuals are vaccinated with this plasmid. The plasmid infects the host cell and expresses the foreign gene or genes and produces the protein inside the cell. The antigen will be expressed on the cell's major histocompatibility complex (MHC) class I molecules, due to the antigen being endogenous. This stimulates the activation of cytotoxic T-cells, which are important for clearance of pathogen-infected cells.

Traditional vaccines enter the MHC class II pathway, due to the fact that the antigens encountered are exogenous. This primarily stimulates antibody responses which are not effective at clearing viruses which are protected by the cell's they reside in. However with DNA vaccines, once the cytotoxic T-cells have been activated, they may lyse infected cells and release the antigen allowing the antibody response to also become stimulated. These are cheaper and easier to store than purifying proteins (very stable) and may also be immunogenic but the side effects remain unclear. The transient production of protein for about a month is enough to evoke a robust immune response. If any of the above vaccines are used, this process is called active immunization. In contrast, passive immunization is an injection of purified antibody to produce rapid but temporary protection. Passive immunization is used to prevent a disease after known exposure, to improve the symptoms of an ongoing disease, to protect immunosuppressed patients, to block the action of bacterial toxins and to prevent disease. Some of the currently available vaccines against bacterial diseases are indicated in (Table 4).

**Table 4** Examples of vaccines available for protection against bacterial diseases.

| Disease | Vaccine components |
|---|---|
| Diphtheria | Inactivated toxin |
| Tetanus | Inactivated toxin |
| Tuberculosis | Attenuated *Mycobacterium bovis* (BCG) |
| Whooping cough | Subcelluar fractions and pertussis toxoid |
| *Haemophilus influenzae* meningitis | Capsular polysaccharide linked to a protein carrier |
| Meningococcal meningitis | Capsular polysaccharide |
| Typhoid | Killed cells of *Salmonella typhi* |
| | Live oral attenuated strain Ty21A |

## 5.2 Antibiotics

Molecules that inhibit the growth of, or kill bacteria are called antibacterial agents. These compounds are normally isolated from microbes and are called antibiotics. There are many millions of antibiotics, produced mainly by soil bacteria and fungi that are active against bacteria but only a few can be used to control or treat human bacterial diseases. These are the few that although toxic to bacteria, but have no significant toxic effects to the human host. The reason for this so called selective toxicity is that the site at which these antibiotics act are either unique to bacteria such as peptidoglycans or very different between prokaryotes and eukaryotes such a ribosomes and nucleic acid synthesis. Table 5 shows the sites in the bacterial cell at which a number of clinically important antibiotics act. Some of the factors that affect antibiotic therapy are listed below:

**Table 5** Target sites for antibiotics in the bacterial cells.

| *Target site* | *Mode of action* | *Example* |
|---|---|---|
| Cell wall (peptidoglycan biosynthesis) | Inhibition of cross-linking | Penicillin (β-lactam antibiotics) |
| | Inhibition of polymerization | Glycopeptide antibiotics (e.g., vancomycin) |
| Protein synthesis | Inhibition of translocation of ribsome | Aminoglycoside antibiotics (e.g., gentamicin) |
| | Inhibition of binding of aminoacyl tRNAs | Tetracycline |
| Nucleic acid synthesis | Inhibition of tetrahydrofolic acid synthesis | Sulphonamides and trimethoprim |
| | Inhibition of DNA gyrase | Quinolone antibiotics (e.g., ciprofloxacin) |

- The antibiotic must reach the site of infection in the host.
- The antibiotic has to reach its target site. This is easier for antibiotics such as penicillin which act on the peptidoglycan than for those like tetracycline which must penetrate through the plasma membrane to reach their target sites, the ribosomes.
- Gram-negative bacteria are often intrinsically resistant to the action of antibiotics due to the presence of the outer membrane which acts as an additional barrier for the antibiotic to cross and protects the peptidoglycan.
- Broad-spectrum antibiotics are effective against a wide range of different Gram-positive bacteria whereas other antibiotics may have only a narrow range.
- All the pathogenic bacteria must be eradicated from the host by either inhibiting the growth of the microbe (bacteriostatic antibiotics), which can then be removed by the immune system or by killing them directly (bactericidal antibiotics).

Antibiotics have proved to be of great benefit to human kind and have ensured that people no longer need to die from diseases such as wound infections. However, recent studies have clearly shown that bacteria can become resistant to the action of antibiotics. The mechanisms by which they do this include: the production of enzymes that break down the antibiotic, reduce the permeability to the antibiotic, and alterations to the target site as shown in Table 6. Antibiotic resistance may arise by mutation but more often the genes for antibiotic resistance are transferred between the bacteria by conjugation, transduction and transformation. Antibiotic resistance carried by plasmids has caused particular concern as these plasmids may carry genes that confer resistance to many different antibiotics at the same time. Multiple-resistant bacteria are therefore becoming a problem particularly in the hospital environment and the fear is that it will not be long before there is a bacterial strain that is untreatable by all known antibiotics. Thus there is a clear need to identify novel targets in bacterial pathogens as well as to seek alternative approaches to develop preventative and therapeutic approaches.

**Table 6** Common mechanisms of antibiotic resistance in bacteria.

| *Mechanism of resistance* | *Mode of action* | *Example* |
|---|---|---|
| Antibiotic inactivation | β-Lactamase | Penicillin resistance |
| | Chloramphenicol acetyl transfease | Chloramphenicol resistance |
| | Aminoglycoside modifying enzymes | Aminoglycoside resistance |
| Reduction in permeability | Reduced uptake | Natural resistance of many Gram-negative bacteria due to presence of outer membrane |
| | Antibiotic efflux | Tetracycline resistance |
| Alteration of target site | Change in target site so that it is no longer sensitive to drug | Sulphonamide resistance |
| | New target site produced that is not sensitive to the antibiotic | Methicillin resistance |
| | Overproduction of target site | Trimethoprim resistance |

## 6. MAJOR BACTERIAL PATHOGENS OF HUMANS

For simplicity, the major bacterial pathogens are divided into several groups as indicated below.

### 6.1 Spirochetes

These are spiral-shaped, motile, Gram-negative bacteria. Their size varies from 0.1 – 3 μm but are very thin and can best be observed by dark-field microscopy or special staining. Spirochetes may be anaerobic, facultative anaerobic or aerobic and include causative agents of human and zoonotic diseases. A zoonosis is an infectious disease transmitted from animals to humans.

### 6.1.1 *Borrelia burgdorferi*

Causative agent of lyme disease. It is initially (within days) characterized by fever, headache, enlarged local lymph nodes, muscle and joint pains and within months may lead to long-lasting arthritis or involve the central nervous system.

**Biology and diagnosis:** Size is 0.2 – 0.5 µm wide and 20 – 30 µm long, spiral-shaped, motile organisms. Although they are Gram-negative but are easily observed using Giemsa staining. Lyme disease is diagnozed based on signs and symptoms such as erythema migranes, facial palsy, or arthritis and a history of possible exposure to infected ticks. Laboratory tests involve immunofluorescence staining or by direct demonstration in blood specimens by dark field microscopy.

**Transmission:** The disease is transmitted by ticks (*Ixodes* spp.). Other mammals such as mice are hosts for ticks and are important for maintaining *B. burgdorferi* in nature.

### 6.1.2 *Borrelia recurrentis*

Causative agent of epidemic relapsing fever. It is characterized by fever, chills, a headache, muscle pain but it is self-limiting.

**Biology and diagnosis:** Size is 0.3 µm wide and 10 – 30 µm long, spiral-shaped, motile organisms. Although they are Gram-negative but are easily observed using Giemsa staining. As the appearance of fever is not a specific diagnostic method, immunofluorescence staining or by direct demonstration of *Borrelia* in blood specimens are recommended.

**Transmission:** Epidemic relapsing fever (etiologic agent–*B. recurrentis*) is transmitted by body lice (*Pediculus* spp.). Endemic relapsing fever (etiologic agents–*B. duttoni, B. hispanica, B. persica, B. hermsii* and *B. turicatae*) is transmitted by ticks (*Ornithodoros* spp.).

### 6.1.3 *Leptospira interrogans*

Causative agent of leptospirosis. It is characterized by fever, chills, headache, muscle pain and may involve the liver, kidney or blood vessels damage but is normally self-limiting.

**Biology and diagnosis:** Size is 0.15 µm wide and 5 – 15 µm long, spiral-shaped, motile organisms that are best observed using dark-field microscopy or immunofluorescence staining.

**Transmission:** Transmission of infections from animals to humans occurs through contact with contaminated soil or moist soils where rodent urine, contaminated tissues or blood are present. *Leptospira* enter humans via skin abrasions or mucous membranes.

### 6.1.4 *Treponema pallidum*

Causative agent of syphilis. Initially (within 2 – 3 weeks), it is characterized by a swollen painless lesion (termed a chancre) at the site of infection and may lead to blisters on the body. If untreated, the disease can progress to secondary and tertiary stages which can occur within 5 – 20 years leading to extensive damage of any tissue involving the cardiovascular system and the CNS.

**Biology and diagnosis:** Size is 0.15 μm wide and 5 – 15 μm long, spiral-shaped, motile organisms that are best observed using dark-field microscopy or immunofluorescence staining.

**Transmission:** Sexually transmitted disease (STD) but can also spread through blood transfusions, mother to foetus, or direct contact with a chancre that may be present on the external genitals, lips and in the mouth.

## 6.2 Aerobic Gram-negative Bacteria

### 6.2.1 *Helicobacter pylori*

Causative agent of stomach ulcers. Although it co-exists with its host but in 1% of the infected individuals develop gastric cancer and 20% develop duodenal ulcers.

**Biology and diagnosis:** Size is 0.5 – 1 μm wide and 2.5 – 5 μm long, motile, spiral, curved or rod-shaped. Diagnosis can be made by the examination of tissue biopsies, Gram staining or using serological tests including enzyme immunoassay.

**Transmission:** Occurs throughout the world with most likely transmission via the foecal-oral or oral-oral route.

### 6.2.2 *Bordetella pertussis*

Causative agent of pertussis (whooping cough). The bacterium grows in the mucosal membranes of the human respiratory tract. It is characterized by violent coughing for 20 – 50 times a day for up to a month.

**Biology and diagnosis:** Size is 0.2 – 0.5 μm wide and 0.5 – 2 μm long, non-motile, cocco-bacillus, cells are present singly or in pairs. Diagnosis can be made by the examination of specimens (nasopharyngeal tissues) together with enzyme immunoassay (EIA).

**Transmission:** Human to human transmission via air borne droplets and highly communicable.

### 6.2.3 *Brucella* spp.

Causative agent of brucellosis. It is characterized by fever, chills lasting for years and may include weight loss, body aches, enlarged lymph nodes, spleen and liver.

**Biology and diagnosis:** Size is 0.5 μm wide and 0.3 – 0.9 μm long, non-motile, cocco-bacillus, cells are present singly or in pairs. Four species of *Brucella* are pathogenic to humans, *B. aborus, B. melitensis, B. suis, B. canis* and their natural hosts are cats, goats/sheep, swines and dogs, respectively. Diagnosis can be made by the examination of biopsies or isolation of the bacterium from the blood, and enzyme immunoassay.

**Transmission:** Through skin abrasions, through the conjunctival sac of the eyes, inhalation of bacteria containing droplets or via contaminated food.

### 6.2.4 *Legionella pneumophila*

Causative agent of Legionnaires' disease (pneumonia-like infection) and Pontiac fever (influenza-like infection). Pontiac fever is self-limiting. Legionnaires' is characterized by fever, chills, muscle pain, coughing, breathing difficulty , and may lead to vomiting, diarrhoea, or involving the CNS resulting in fatal consequences.

**Biology and diagnosis:** Size is 0.3 – 0.9 μm wide and 2 – 20 μm long, motile, rod-shaped. Diagnosis can be made by isolation of the organisms or by direct immunofluorescence testing or enzyme immunoassay.

**Transmission:** Transmission is via airborne droplets or waterborne.

### 6.2.5 *Neisseria gonorrhoeae*

Causative agent of gonorrhoea. In males, it is characterized by a pus-like discharge, painful urination and may even result in complete closure of the urethra (the canal that connects the bladder to the outside). In females, there is rarely urethral discharge but rather dysuria, painful urination and inflammation of the uterine tubes and are major causes of infertility in women. Endocervicitis may be present as slight vaginal pus discharge. Pelvic inflammatory disease may occur and disseminate throughout the reproductive tract. In rare cases, bacteria may disseminate via the bloodstream to produce a rash and arthritis. Babies born to infected mothers may develop eye infections (gonococcal ophthalmia/ophthalmia neonaturum), which can result in blindness, if eye drops/ointment are not administered at birth.

**Biology and diagnosis:** Size is 0.6 – 1 μm in diameter, non-motile, diplococci. Diagnosis can be made by isolation of the organisms or by immunofluorescence testing or enzyme immunoassay.

**Transmission:** Sexually transmitted disease (STD) but can also spread by the mother to the foetus when it passes through the infected birth canal.

### 6.2.6 *Neisseria meningitidis*

A major causative agent of bacterial meningitis (also known as meningococcal meningitis). The organism spreads from the nasopharynx to the bloodstream and finally invades the CNS to produce disease. While in the bloodstream, the bacterium may grow in large numbers to produce bacteremia causing severe damage to the blood vessels resulting in hemorrhages (noticeable rash on the skin). The disease is characterized by fever, stiff neck, vomiting, severe headache, coma and death within a few hours.

**Biology and diagnosis:** Size is 0.6 – 1 μm in diameter, non-motile, diplococci. Diagnosis can be made by direct examination of the bacteria and serological test using latex agglutination for the detection of capsular polysaccharides or using enzyme immunoassay.

**Transmission:** Human to human via respiratory droplets.

### 6.2.7 *Pseudomonas aeruginosa*

Causative agent of lung infection, eye infection and skin infection. It is an opportunistic pathogen and colonizes the injured epithelial cell surfaces, resulting in serious consequences.

**Biology and diagnosis:** Size is 0.5 – 1 μm wide and 1.5 – 5 μm long, motile, rod-shaped, cells are present singly or in pairs. Diagnosis can be made by direct examination of the bacteria or immunoassay.

**Transmission:** Exposure of injured epithelial cell surface to the contaminated surfaces, of which water is the most common form of transmission.

Other aerobic Gram-negative bacteria which may cause human infections are *Moraxella* spp. *Francisella* spp. and *Alcaligenes* spp.

## 6.3 Facultative Anaerobic Gram-negative Bacteria

### 6.3.1 *Calymmatobacterium granulomatis*

Causative agent of granuloma inguinale with the skin lesions and subcutaneous tissue of genital and anal regions.

**Biology and diagnosis:** Size is 0.5 – 1.5 μm wide and 1 – 2 μm long, rod-shaped. Diagnosis can be made by direct examination of bacteria, Giemsa but not Gram staining is effective.

**Transmission:** May be sexually transmitted (although this claim is controversial).

### 6.3.2 *Escherichia coli*

Causative agent of the gastrointestinal and the CNS infections. The gastrointestinal infections are associated with fever, diarrhoea which may contain blood, vomiting within days and may result in severe dehydration, shock and death. *E. coli* meningitis is observed in neonates and is characterized by fever, a stiff neck, vomiting, severe headache, coma and death.

**Biology and diagnosis:** Size is 1 – 1.5 μm wide and 2 – 6 μm long, motile, rod-shaped, cells are present singly or in pairs. Diagnosis can be made by isolation of the organisms, Gram staining or by immunofluorescence testing.

**Transmission:** Through contaminated food / water.

### 6.3.3 *Gardnerella vaginalis*

Causative agent of vaginosis (foul-smelling vaginal discharge).

**Biology and diagnosis:** Size is 0.5 μm wide and 1.5 – 2.5 μm long, non-motile, rod-shaped. Diagnosis can be made by examination of organisms in the vaginal secretions and growing on selective agar.

**Transmission:** Sexually transmitted disease.

### 6.3.4 *Haemophilus influenzae*

Causative agent of pneumonia, bronchitis, meningitis. It is characterized by a sore throat, fever, breathing problems, invasion of the bloodstream, and death.

**Biology and diagnosis:** Size is less than 0.3 – 0.8 μm in diameter, non-motile, rod-shaped. Diagnosis can be made by isolation and direct examination of the organism in the clinical specimens or by serological testing.

**Transmission:** Human to human via respiratory droplets.

### 6.3.5 *Pasteurella multocida*

Causative agent of cellulites (infection of the sub-cutaneous) tissue.

**Biology and diagnosis:** Size is 0.3 – 1 μm wide and 1 – 2 μm long, non-motile, ovoid or rod-shaped, cells are present singly or in pairs. Diagnosis can be made by culture of pus taken from the lesions.

**Transmission:** Normally present in the respiratory tract of dogs and cats. Transmitted to humans via a bite of a domestic animal.

### 6.3.6 *Proteus vulgaris / P. mirabilis*

Causative agent of gastrointestinal, urinary tract and extraintestinal infections.

**Biology and diagnosis:** Size is 0.4 – 0.8 µm wide and 1 – 3 µm long, motile, rod-shaped, cells are present singly or in pairs. Diagnosis can be made by culture and biochemical tests.

**Transmission:** Through contaminated food/water.

### 6.3.7 *Salmonella* spp.

Some species can cause gastroenteritis (inflammation of stomach and intestines) and may result in enteric fever. *Salmonella typhi* is the causative agent of typhoid fever and enteric fever. Gastroenteritis is associated with nausea, vomiting and diarrhoea. In typhoid fever, the bacterium penetrates the intestine, enters local lymph nodes and gains access to the bloodstream and disseminates to various organs. Typhoid is characterized by fever, chills, aches, weakness, and constipation, rather than diarrhoea, leading to bacteremia and infection of other organs such as the CNS. Of interest, enteric fever typically involves the liver and/or spleen with prolonged fever.

**Biology and diagnosis:** Size is 0.5 – 1.5 µm wide and 2 – 5 µm long, motile, rod-shaped. Diagnosis for gastroenteritis is made by culture from stools and blood.

**Transmission:** Through contaminated, under cooked food (meat and dairy products are typical) or untreated water.

### 6.3.8 *Shigella* spp.

Causative agent of shigellosis, i.e., bacterial dysentery (diarrhoea with blood and mucus). It is characterized by severe abdominal pain and cramps but it is self-limiting within days.

**Biology and diagnosis:** Size is 0.5 – 0.8 µm wide and 2 – 3 µm long, non-motile, rod-shaped, cells are present singly or in pairs. Diagnosis can be made by culture from stools and by using biochemical tests.

**Transmission:** Through contaminated food/water, foecal-oral route.

### 6.3.9 *Streptobacillus moniliformis*

Causes fever, chills, vomiting, headache, rash and arthritis.

**Biology and diagnosis:** Size is 0.1 – 0.8 µm wide and 1 – 5 µm long, non-motile, rod-shaped, present in long chains. Diagnosis can be made by culturing from clinical specimens and by using biochemical tests.

**Transmission:** Through bites of infected mice, rats, cats or by ingesting food/water contaminated with rat faeces.

### 6.3.10 *Vibrio cholerae*

Causative agent of cholera. It is characterized by vomiting, abdominal pain, excessive amounts of watery dehydrating diarrhoea.

**Biology and diagnosis:** Size is 0.5 – 0.8 μm wide and 1.3 – 2.5 μm long, motile, rod or curved-shaped. Diagnosis can be made by culturing from stool specimens and serological tests.

**Transmission:** Through contaminated water/food.

### 6.3.11 *Yersinia pestis*

Causative agent of plague. There are three major clinical forms, bubonic plague, septicemic plague and pneumonic plague. The former is characterized by fever and appearance of bubo (swollen local region in the groin or armpit), and if not treated leads to bacteremia (septicemic plague), septic shock and can spread to the lungs leading to pneumonic plague. Pneumonic plague is characterized by fever, breathing difficulty, and death in some patients.

**Biology and diagnosis:** Size is 0.5 – 0.8 μm wide and 1 – 3 μm long, non-motile, rod-shaped. Diagnosis can be made by culturing from clinical specimens and Gram staining.

**Transmission:** Bubonic plague is transmitted by the fleas of urban and sylvatic rodents mainly (bites or ingestion) to humans; and pneumonics plague is airborne, i.e., person-to-person.

### 6.3.12 *Yersinia enterocolitica*

Causative agent of enterocolitis (usually in children). It is characterized by fever, abdominal pain and diarrhoea but is self-limiting.

**Biology and diagnosis:** Size is 0.5 – 0.8 μm wide and 1 – 3 μm long, non-motile, rod-shaped. Diagnosis can be made by examining stool specimens.

**Transmission:** Through ingestion of contaminated food / water.

Other facultative anaerobic Gram-negative bacteria which may cause human infections are *Serratia* spp. *Klebsiella* spp. *Enterobacter* spp. *Morganella* spp. and *Citrobacter* spp.

## 6.4 Anaerobic Gram-negative Bacteria

### 6.4.1 *Prevotella* spp.

Found in infected oral cavities but their clinical significance is not clearly understood.

**Biology and diagnosis:** Size is 0.5 – 0.8 μm wide and 1 – 3 μm long, non-motile, rod-shaped. Diagnosis can be made by culturing from the clinical specimens.

**Transmission:** Unknown but likely to be through human to human contact.

Other anaerobic Gram-negative bacteria which may cause human infections are *Porphyromonas* spp. (*P. gingivalis, P. endodontalis*).

## 6.5 Rickettsia and Chlamydia

Obligate, intracellular bacteria that live and multiply within their eukaryotic host cells and cause zoonotic diseases.

### *6.5.1* ***Rickettsia*** **spp.**

*Rickettsia rickettsii* is the causative agent of Rocky Mountain spotted fever and *R. prowazekii* is the causative agent of epidemic typhus fever with life-threatening consequences.

**Biology and diagnosis:** Size is 0.2 µm wide and 0.5 – 1.3 µm long, rod-shaped, obligate intracellular organisms. Diagnosis can be made by serological tests or by culturing in host cell cultures.

**Transmission:** Through bite of infected tick/mite or through faeces of infected lice/flea.

### *6.5.2* ***Coxiella burnetii***

Causative agent of Q-fever. It is characterized by flu-like symptoms, fever, headaches, pneumonia and hepatitis.

**Biology and diagnosis:** Size is 0.2 µm wide and 0.5 – 1.3 µm long, cocco-bacillus, obligate intracellular organisms. Diagnosis can be made by serological tests or by culturing in host cell cultures.

**Transmission:** Through inhalation of contaminated droplets and associated with domestic animals.

### *6.5.3* ***Ehrlichia*** **spp.**

Causative agent of Ehrlichiosis. Several species of *Ehrlichia* (such as *Ehrlichia chaffeensis*) can cause this disease with symptoms such as fever, headache, fatigue, muscle aches, nausea, vomiting, diarrhoea, cough and joint pains. Symptoms normally appear after an incubation period of 5-10 days following the tick bite.

**Biology and diagnosis:** Size is 0.5 – 1.6 µm long, variable shapes, obligate intracellular organisms. Diagnosis can be made by Giemsa staining and serological tests.

**Transmission:** Transmitted by tick bites.

### *6.5.4* ***Bartonella*** **spp.**

Causes fever, malaise, hepatitis, bacteremia, mostly limited to immunocompromised patients.

**Biology and diagnosis:** Size is 0.5 µm wide and 1 – 2 µm long, non-motile, rod-shaped. As these organisms can be cultured, thus they are not obligate intracellular bacteria. Diagnosis can be made by culturing on selective agar and serological tests.

**Transmission:** Transmitted by domesticated cats, cat flea, body louse, and ticks (*Ixodes* spp. and *Dermacentor* spp.).

### 6.5.5 *Chlamydia trachomatis*

Causes trachoma (inner eye lid infection and may include the cornea), conjunctivitis and genital infections (major cause of non-gonococcal urethritis in males). May also cause corneal infections.

**Biology and diagnosis:** Size is 0.3 – 0.9 µm in diameter, cocci-shaped, obligate intracellular organisms. They have two stages in their life cycle, an infectious, rigid, metabolically inactive stage (elementary body) and a delicate, metabolically active stage (reticulate body). Diagnosis can be made by serological, immunfluorescence tests and by using nucleic acid probes.

**Transmission:** Through exposure to contaminated fingers, sexual contact, inanimate objects and flies.

### 6.5.6 *Chlamydia pneumoniae*

Causes sore throat, bronchitis, and pneumonia characterized by prolonged coughing and asthma by damaging epithelial lining of the respiratory tract.

**Biology and diagnosis:** Size is 0.3 – 0.9 µm in diameter, cocci-shaped, obligate intracellular organisms. They have two stages in their life cycle, an infectious, rigid, metabolically inactive stage (elementary body) and a delicate, metabolically active stage (reticulate body). Diagnosis can be made by serological, tests and by using nucleic acid probes.

**Transmission:** Through inhalation of contaminated droplets.

## 6.6 Mycoplasmas

### 6.6.1 *Mycoplasma pneumoniae*

Causes infections of the respiratory tract (normally in 5 – 9 year old children). It is characterized by prolonged coughing, fever and headaches.

**Biology and diagnosis:** Size is 0.1 – 0.3 µm in diameter. Diagnosis can be made by culturing from specimens and serological tests.

**Transmission:** Through inhalation of contaminated droplets, close personal contact.

Other *Mycoplasma* which may cause human infections are *Ureaplasma* spp. that causes genital infections in humans.

## 6.7 Gram-positive Bacteria

### 6.7.1 *Enterococcus faecalis*

Causes urinary tract infections.

**Biology and diagnosis:** Size is 0.6 – 2.5 μm long, non-motile, cocci-shaped, cells are present in pairs or in short chains. Diagnosis can be made by isolation and biochemical tests.

**Transmission:** Through inhalation of contaminated droplets, or through contaminated food / water.

### 6.7.2 *Staphylococcus aureus*

Causes various infections including skin lesions (boils and carbuncles), pneumonia, meningitis, endocarditis (heart) and toxic shock syndrome. The exotoxin but not the bacteria cause food poisoning.

**Biology and diagnosis:** Size is 0.5 – 1.5 μm in diameter, non-motile, cocci-shaped, cells are present singly, in pairs or in grape-like clusters (Staphylococcus is a Greek word meaning bunch of grapes). They are non-motile, anaerobic organisms. Diagnosis can be made by biochemical tests and serological tests.

**Transmission:** Through inhalation of contaminated droplets, or through contaminated food/water and insects.

### 6.7.3 *Streptococcus*

Bacteria in this genus are non-motile, facultative anaerobes and consist of an enormous number of species, responsible for a variety of infections involving various tissues and organs. *Streptococcus* that are major human pathogens are divided into two groups.

**6.7.3i Group A streptococci (GAS)** Most commonly associated with human diseases and include *Streptococcus pyogenes* and *S. pneumoniae*. The former is responsible for pharyngitis (strep throat), fever, acute rheumatic fever, sepsis and shock. The latter (pneumococci) are responsible for pneumonia, sinusitis, bacteremia and meningitis.

**Biology and diagnosis:** Size is 0.5 – 2 μm in diameter, non-motile, cocci-shaped, cells are present in pairs or in short chains. Diagnosis can be made by isolation, Gram staining, biochemical and serological tests.

**Transmission:** Through inhalation of contaminated droplets, or through contaminated food / water.

**6.7.3ii Group B streptococci (GBS)** Cause neonatal pneumonia, sepsis and meningitis.

**Biology and diagnosis:** Size is 0.5 – 2 μm in diameter, non-motile, cocci-shaped, cells are present in pairs or in short chains. Diagnosis can be made by isolation, Gram staining, biochemical and serological tests.

**Transmission:** Mother to foetus during birth.

### 6.7.4 *Bacillus anthracis*

Causative agent of cutaneous and inhalational anthrax.

**Biology and diagnosis:** Size is 1 – 1.5 μm wide and 3 – 10 μm long, rod/ovoid-shaped and form spores. Diagnosis can be made by demonstrating bacteria in the specimen, Gram staining and serological tests.

**Transmission:** Through inhalation of contaminated droplets, or by contact with infected animals.

### 6.7.5 *Clostridium botulinum*

Causative agent of botulism (paralysis). Botulism is not an infectious disease but intoxification caused by bacterial toxins, i.e., botulinum toxin and bacteria may be completely absent. The bacterium is divided into 8 types (A, B, C alpha, C beta, D, E, F, G) on the basis of specific exotoxins produced, all of which form spores.

**Biology and diagnosis:** Size is 0.3 – 2 μm wide and 1 – 2 μm long, rod-shaped, cells are present in pairs or in short chains. Diagnosis is based on symptoms and treatment is with specific antitoxins.

**Transmission:** The spores are present everywhere in nature and usually germinate when they are in an anaerobic environment such as home canned beans, corn or other low acid food. The growing cells release a neurotoxin (the most powerful known to man). It is a toxin which if ingested before it is heat inactivated by cooking which is the killer. This toxin is notorious and may be used as a bioterror agent (also see Chapter 2).

Other species of *Clostridium* are *C. difficile*, which causes colitis (colon tissue destruction), *C. perfringens*, which causes gas gangrene and *C. tetani*, which causes tetanus or lockjaw.

### 6.7.6 *Listeria monocytogenes*

Causative agent of listeriosis. Usually limited to immunocompromised patients or pregnant women and can lead to bacteremia, abortion, meningitis and meningo-encephalitis.

**Biology and diagnosis:** Size is 0.4 μm wide and 0.5 – 2 μm long, motile, rod-shaped, cells are present singly or in short chains. Diagnosis can be made by cultures of clinical specimens, Gram staining and serological tests.

**Transmission:** Through ingestion of contaminated food.

### 6.7.7 *Erysipelothrix rhusiopathiae*

Causes zoonotic infections (from animals to humans). Human infections are primarily limited to individuals who handle fish or pigs. Clinical categories include (i) a localized cutaneous form (most common) associated with a throbbing itching pain and swelling of the finger or part of hand, (ii) a generalized cutaneous form, and (iii) a septicemic form (associated with the heart disease endocarditis).

**Biology and diagnosis:** Size is 0.2 – 0.4 μm wide and 0.8 – 2.5 μm long, non-motile, Gram-positive, nonsporulating rod and is a facultative anaerobe. Diagnosis is made by the isolation of the organisms from tissue biopsies or blood.

**Transmission:** The organism is likely to be found in faecally contaminated environments and enters into humans through scratches or lesions on the surface of the skin.

### 6.7.8 *Actinomyces israelii*

Causative agent of actinomycosis. Bacteria are normally present in the mouth but can cause infection following injury such as a broken jaw, tooth extractions etc, which introduce them to the deeper tissues.

**Biology and diagnosis:** Size is 0.2 – 1 μm wide and 2 – 5 μm long, non-motile, rod-shaped, cells are present singly, in pairs or short chains forming distinct V or Y forms. Diagnosis can be made by cultures of clinical specimens, and staining.

**Transmission:** Injury to mucosal surfaces.

### 6.7.9 *Corynebacterium diphtheriae*

Causative agent of diphtheria. It is associated with respiratory tract infection by the bacterium and systemic effects of its exotoxin. The disease is characterized by a sore throat often with a thick adherent membrane composed of dead cells, bacteria, phagocytes and fibrin, headache, nausea and death can occur from heart tissue damage by the exotoxin.

**Biology and diagnosis:** Size is 0.3 – 0.8 μm wide and 1.5 – 8 μm long, non-motile, rod-shaped, cells are present singly, in pairs or short chains forming distinct V forms. Diagnosis can be made by clinical symptoms, isolation of the organism, and demonstration of its toxin.

**Transmission:** Through inhalation of contaminated airborne droplets, or by direct contact with skin infections.

Other Gram-positive bacteria which may cause human infections are *Aerococcus viridans*, *Micrococcus luteus*, and *Sarcina* spp. that may cause pulmonary, skin, gastrointestinal and sub-cutaneous infections.

## 6.8 Mycobacteria

### *6.8.1 Mycobacterium tuberculosis*

Causative agent of tuberculosis. It is infection of the respiratory system characterized by a cough, fever and weight loss. Bacteria avoid host defences by forming lesion (tubercle). Tissue injury results from cell-mediated hypersensitivity.

**Biology and diagnosis:** Size is 0.2 – 0.7 µm wide and 1 – 10 µm long, non-motile, rod-shaped. Diagnosis can be made by clinical symptoms (chest radiograph), and demonstration of the organism in the sputum or other specimens (acid fast staining).

**Transmission:** Through inhalation of contaminated airborne droplets or by direct contact with infected patients.

### *6.8.2 Mycobacterium leprae*

Causative agent of leprosy. It has two forms, tuberculoid leprosy and lepramatous leprosy. The former is associated with dry, pale patches on any surface of the body and may develop into loss of nerve sensation due to bacterial invasion of peripheral sensory nerves. The latter is associated with extensive skin involvement and damage can result in the loss of facial bones, fingers and toes.

**Biology and diagnosis:** Size is 0.2 – 0.7 µm wide and 1 – 10 µm long, non-motile, rod-shaped. Diagnosis can be made by clinical symptoms and demonstration of the organism in the infected tissue (acid fast staining).

**Transmission:** Through inhalation of contaminated airborne droplets, or by direct contact with infected patients.

Other species of *Mycobacterium* are *M. avium*, which causes pulmonary diseases, infection of the skin, bones and joints.

# 7. *ESCHERICHIA COLI* AS A MODEL BACTERIUM

*Escherichia coli* was first described by Dr. Theodor Escherich, in the late 19th century as a normal microbial inhabitant of the healthy individuals. Now *E. coli* is recognized as a Gram-negative, short rod, facultatively anaerobe, which is a universal inhabitant of the intestinal tract of animals (Fig. 1). In humans, this colonization starts to happen immediately after birth and they are commonly found in the bowels of neonates. Thus the human association with *E. coli* is life long and for the most part trouble-free. However, some strains of *E. coli* are pathogenic and cause serious and fatal human infections (Fig. 22). *Escherichia coli* strains producing a diverse array of infections are classified by serotypes. Serotyping is based on three antigenic determinants:

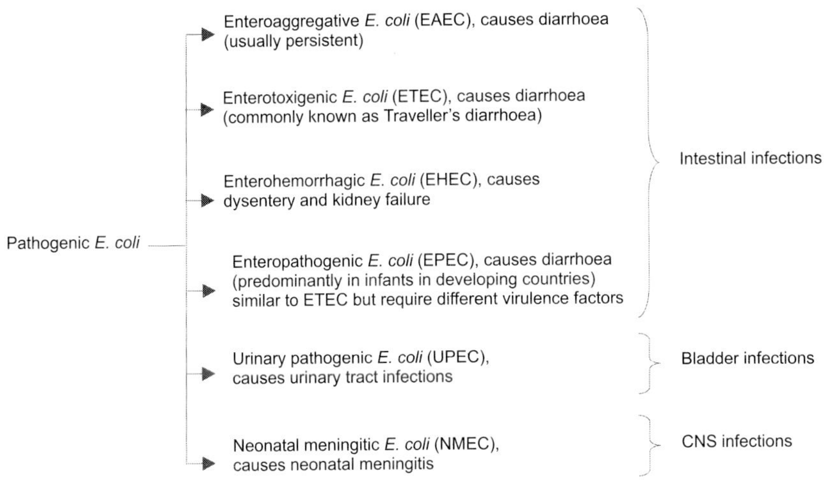


**Fig. 22** Pathogenic *E. coli* strains.

(i) lipopolysaccharide (O)
(ii) capsule (K) and
(iii) flagella (H)

There are more than 700 serotypes but certain *E. coli* serotypes are more abundant in specific diseases. For example, the notorious EHEC O157:H7 commonly cause dysentery and NMEC O18:K1:H7 commonly cause neonatal meningitis. Here, NMEC is represented as a model organism and our current understanding of their pathogenesis and pathophysiology associated with meningitis is described.

## 7.1 Neonatal Meningitis

*Escehrichia coli* is a major cause of neonatal meningitis. The incidence rate of *E. coli* neonatal meningitis is estimated at 1.2 per 10,000 live births with a mortality rate ranging from 20 to 40%. It principally affects infants (< 30 days old), with premature and special care babies particularly at risk, with a high risk of either death or long term disability. A high level of mortality is partly a result of the rapidity of the disease onset requiring immediate diagnosis and treatment. In addition, the common association with septicemia (presence of bacteria in blood), itself has serious consequences such as high fever, headache, nausea, and hypotension i.e., low blood pressure due to inadequate blood flow to the heart, the brain and other organs (also known as shock), resulting in death. Septicemia can develop quickly with the appearance of a rash under the skin (representative of bleeding under the skin). If untreated, spots/bruises get bigger and become multiple areas of bleeding under the skin and can appear anywhere on the body. In contrast, meningitis is inflammation of the meninges (membranes enclosing the brain). Normally clear cerebrospinal fluid (CSF) becomes cloudy with leukocytes (white blood cells) attracted to the site of inflammation. The swelling around the brain increases pressure inside the skull causing an intense headache. Inflammation of spinal meninges affect nearby muscles causing a stiff neck. In addition, the brain function is affected as meninges inflammation can diminish flow of blood to the brain. If not diagnozed early and treated aggressively, meningitis results in death within hours to days. Even with treatment, up to 50% survivors may sustain developmental disability such as hearing loss, learning disability, loss of vision and other complications.

## 7.2 Pathogenesis of *E. coli* K1 Meningitis

Among various serotypes of NMEC, *E. coli* strains expressing K1 capsule are significantly associated with meningitis. In fact, more than 80% of NMEC cases are due to *E. coli* K1. Much of our understanding of the pathogenesis and pathophysiology of *E. coli* meningitis comes from the K1 strains. The basic pathogenic mechanisms associated with *E. coli* K1 meningitis are highly complex and remain unclear. However, based on our current understanding, they can most likely be divided into the following steps (Fig. 23):

(i) Colonization of the intestinal tract.
(ii) Crossing the intestinal lumen to gain entry into the bloodstream.
(iii) Survive host defence mechanisms.
(iv) Magnitude of bacteremia (i.e., a threshold level).
(v) Bacterial invasion of the brain microvascular endothelial cells (blood-brain barrier).
(vi) Traversal of the blood-brain barrier as live bacteria and entry into the CNS.
(vii) Survival and replication in the subarachnoid space, producing disease.

The sequence of events initiates with *E. coli* acquisition from the mother's flora or from the environment. *E. coli* colonizes the infant's intestinal tract and translocate from the intestinal lumen to the bloodstream. This is the first critical step but the precise mechanisms remain unclear. It is shown that bacterial translocation of the intestinal lumen requires an *E. coli* cell density of at least $10^8$ per gram of faeces. However, the urinary tract may also be a route of entry into the bloodstream (in up to 20% cases). This is followed by a sustained high-level bacteremia. This is due to the ability of bacteria to survive intravascularly by evading and/or resisting host defence

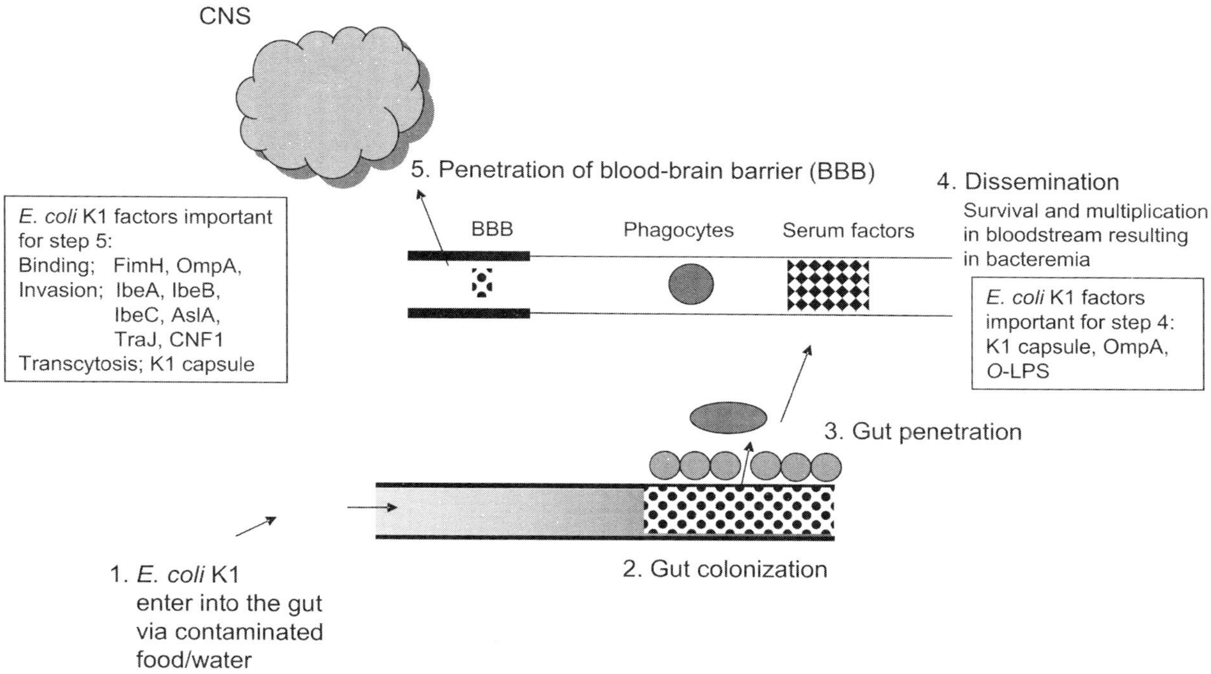


**Fig. 23** Mechanisms of *E. coli* K1 neonatal meningitis.

mechanisms and multiplication. Indeed, bacteremia is the second critical step in *E. coli* meningitis. For example, bacteremia below or equal to $10^3$ colony forming units per milliliter leads to meningitis in 5% of cases, while bacteremia above $10^3$ colony forming units per milliliter leads to meningitis in more than 50% of cases. However, the mechanisms of *E. coli* K1 to survive and multiply in the bloodstream are highly complex. To this end, recent studies have shown that *E. coli* K1 can survive within the macrophages and inhibit neutrophil NADPH-oxidase to produce reactive oxygen species intracellularly. The ability of *E. coli* K1 to survive the macrophage onslaught and/or to evade other host immune responses is critical to achieve a high level of bacteremia. A limited number of findings suggest that an outer membrane protein A (OmpA) and K1 capsule are important *E. coli* determinants and are required to induce a high level of bacteremia. The OmpA is a major outer membrane protein (approximate molecular weight of 35 kDa) in *E. coli*. *E. coli* K1 strains lacking OmpA show significantly reduced bacteremia in a neonatal rat model of *E. coli* meningitis. In addition, K1 capsule and *O*-lipopolysaccharide (*O*-LPS) are important bacterial determinants in *E. coli* K1 for induction of a high level of bacteremia. This is due to their abilities to protect bacteria from complement-mediated lysis as well as protection from opsonization-mediated phagocytic killing. Complement is activated via either the classical or the alternative pathway. It is believed that the *O*-LPS is important in *E. coli* resistance to the classical pathway, while the K1 capsule may be involved in resistance to the alternative pathway but the precise mechanisms and additional responsible factors remain unclear.

Once a high level of bacteremia is achieved, *E. coli* K1 is then able to cross the blood-brain barrier, enter the brain, setting up an inflammatory response in the meningeal membrane thus causing meningitis. The mechanism by which the blood-brain barrier is breached is the area of interest, not only in the search for novel treatments for meningitis, but to better understand the blood-brain barrier to create therapies that can target the CNS more efficiently while designing others that will not cross the blood-brain barrier so limiting possible side effects.

The blood-brain barrier in humans is a monolayer of endothelial cells joined by tight junctions which form the blood vessel of the fine capillaries (microvessels) and supply the brain with nutrients. The blood-brain barrier regulates the exchange of molecules with the brain and protects it against pathogens as well as toxins. It is proposed that the process by which *E. coli* K1 crosses the blood-brain barrier is a stepwise sequence of events and includes: adhesion to the host cell membrane, cell invasion, transfer through the cytoplasm and exiting the cell as live bacteria to invade the CNS. The entire process requires the function of various bacterial determinants and host molecules and is termed bacterial translocation/crossing of the blood-brain barrier. Most of the work with regard to these considerations has been carried out over the last 10 years and is in its infancy.

### 7.2.1 Adhesion

To cross the blood-brain barrier, bacteria have first to overcome the forces imposed by the blood flow allowing them to move out of circulation. To do this they must adhere to the brain endothelial cells, which constitute the blood-brain barrier. It makes sense that external components of the bacteria will come into contact with the host cell first, suggesting them as likely candidates for a role in the adhesion process. Fimbriae, the most external of all, are classified according to the antigenic properties, type 1 fimbriae and S fimbriae. Type 1 but not S-fimbriae are implicated in *E. coli* binding to the endothelial cells. *E. coli* typically expresses 100 – 300 fimbriae, each approximately 0.2-2.0 μm long and 0.7 nm wide. They are heteropolymers based around a 17 kDa structural protein FimA and express a 29 kDa protein FimH at their tips. FimH is a lectin-like protein with a high affinity to mannose residues. It was additionally observed that mannose inhibited *E. coli* binding to the endothelial cells suggesting a role for type 1 fimbriae. A feature of interest, and with possible virulence implications, is the ability of *E. coli* to determine whether type 1 fimbriae are expressed. It is a phase, dependent on the 'all or nothing' expression. This is a result of inversion or switching 'on' or 'off' of the promoter for the *fim*A gene coding for the fimbrial component FimA by upstream genes *fim*B and *fim*E. It is shown that *E. coli* K1 adherance to the endothelial cells is reduced in the non-fimbriated state and that increased levels of *fim* gene expression are associated with fimbriated *E. coli* K1 association with the brain microvascular endothelial cells (BMEC), tending to confirm the binding role of type 1 fimbriae. Highlighting the refinement of bacterial-host interaction, it has been shown that using a guinea pig erythrocytes as target cells that type 1 fimbriae expressing FimH are able to modulate the level of attachment to the receptor such that it is augmented by an increase in sheer force applied by a flowing liquid medium. This feature would certainly be of advantage to an invasive strain of *E. coli*, causing the initial host-pathogen interactions to be more effective and allowing other determinants to bind, resulting in more intimate contact for more effective interactions and subsequent invasion.

Among other determinants, OmpA has been shown to be important in *E. coli* K1 binding to human BMEC (HBMEC). OmpA is a 35 kDa molecule and is the most abundant protein in the outer membrane. Structurally it's N terminal forms an anti-parallel β-barrel with 8 transmembrane strands resulting in 4 extracellular hydrophilic loops. It was subsequently shown that OmpA is involved in the HBMEC binding through N-glucosamine (GlcNAc) epitopes of HBMEC glycoprotein gp96 (96 kDa) expressed on the host cell membrane. This was confirmed with the finding that the chitooligomers (GlcNAc1, 4-GlcNAc oligomers) block the *E. coli* K1 binding and invasion of HBMEC *in vitro* and traversal of the blood-brain barrier *in vivo*.

### 7.2.2 *Invasion*

Following initial binding, *E. coli* must invade the host cell membrane to gain entry into the host cell. Both type 1 fimbriae and OmpA contribute to the invasion of HBMEC as a result of their effects on the HBMEC binding. In addition, several bacterial determinants responsible for *E. coli* K1 invasion of the HBMEC have been identified and include Ibe proteins (IbeA, IbeB, IbeC), AslA, TraJ, and an A-B type toxin, cytotoxic necrotizing factor 1 (CNF1). The functional involvement of these *E. coli* determinants in *E. coli* K1 invasion of HBMEC *in vitro* and traversal of the blood-brain barrier *in vivo* have been shown with deletion and complementation experiments. Moreover, a 37 kDa laminin receptor precursor (LRP) has been identified as HBMEC receptor for bacterial CNF1 protein (A-B toxin with approximate molecular weight of 110 kDa). LRP is a ribosome-associated cytoplasmic protein shown to be a precursor of 67 kDa laminin receptor (LR). Upon maturation, LR is recruited to the cell membrane and acquires the ability to bind to the beta-chain of laminin with high affinity. Recent studies have shown that incubation of HBMEC with *E. coli* K1 upregulates 67LR expression by HBMEC and recruits 67LR to the invading *E. coli* K1 in a CNF1-dependent manner. Overall, these findings suggest that *E. coli* K1 invasion of HBMEC is a complex process that requires the function of various bacterial proteins as well as an A-B toxin.

### 7.2.3 *Host cell cytoskeletal rearrangement*

*E. coli* K1 invasion of HBMEC requires rearrangements of the actin cytoskeleton and blocking actin condensation with microfilament-disrupting agents such as cytochalasin D abolishes *E. coli* K1 invasion of HBMEC. Also, transmission and scanning electron microscopy show that *E. coli* K1 induces membrane protrusions at the entry site on the HBMEC surface, confirming the involvement of the cytoskeleton. Thus, much of the work has focussed on *E. coli* and host factors involved in HBMEC cytoskeletal changes. The indication that invading *E. coli* K1 are in some way manipulating host cell cytoskeletal arrangements is shown with the finding that a protein tyrosine kinase (PTK) inhibitor, genistein blocked invasion, while it did not reduce adhesion, demonstrating the role of PTK and a distinction between adhesion and invasion. Subsequently, the HBMEC signalling molecules such as focal adhesion kinase (FAK) and its associated cytoskeletal proteins paxillin, phosphatidylinositol 3-kinase (PI3K), and cytosolic phopholipase A2 (cPLA$_2$) are innvolved in *E. coli* K1 invasion of HBMEC, most likely through their effects on actin cytoskeleton rearrangements. The relationship between FAK and PI3K in HBMEC was further confirmed with the finding that FAK operates upstream of PI3K in the signalling pathway. The tyrosine phosphorylation of FAK, paxillin and PI3K are mediated by *E. coli* K1 determinants such as OmpA.

Another *E. coli* K1 virulence facter, cytotoxic necrotising factor 1 (CNF1), was evaluated for a role in the invasion and penetration of the HBMEC. CNF1 is a AB type bacteria toxin (see chapter 4) that contributes to *E. coli* K1 invasion of HBMEC *in vitro* and traversal of the blood-brain barrier *in vivo* through its activation of RhoGTPases leading to cytoskeletal rearrangements. Rho GTPases are the key regulators of actin cytoskeleton in all eukaryotic cells and link external signals to the cytoskeleton by switching a GDP-bound inactive Rho to a GTP-bound active Rho resulting in specific cytoskeletal rearrangements. These involve three major intracellular pathways including i) RhoA pathway leading to stress fibre formation, ii) Rac1 activation triggers lamellipodia formation, and iii) Cdc42 activation promotes filopodia formation. The stress fibre formation plays an important role in *E. coli* K1 entry into the HBMEC suggesting the involvement of RhoA. In support, CNF1 is shown to activate RhoGTPases, such as RhoA, Cdc42 but not Rac1, by a site-specific deamidation of glutamine 63 of RhoA (or glutamine 61 of Cdc42). CNF1-mediated RhoA activation is correlated with the levels of LRP expression in HBMEC further confirming LRP as a receptor for CNF1. RhoA activation leads to ezrin activation (i.e., phosphorylation). Ezrin is a member of ERM (ezrin, radixin and moesin) protein family, which connects F-actin filaments to plasma membranes and induces formation of membrane protrusions responsible for *E. coli* K1 internalization of HBMEC. Overall, CNF1-mediated cytoskeletal changes involve RhoGTPases and these pathways are distinct from OmpA-mediated cytoskeletal changes, which involve FAK, paxillin and PI3K (Fig. 24).

### 7.2.4 Traversal of the blood-brain barrier as live bacteria

Once invaded in the HBMEC, *E. coli* K1 remains in an enclosed vacuole, transmigrates through the HBMEC and exits from the other side as live bacteria to produce disease (Fig. 25). During the transportation in an enclosed vacuole, *E. coli* do not multiply but remain viable, which in the normal course of events would be killed by lysosomal enzymes released into the vacuole by fusion with lysosomes (Fig. 25). It is not clear what bacterial factors are important for these steps and their precise mechanisms remain unidentified. At least one of the factors seems to be K1 capsule. For example, *E. coli* $K1^+$ but not $K1^-$ strains can escape lysosomal fusion. This is shown with findings that vacuoles containing *E. coli* $K1^+$ recruit early endosomal markers such as early endosomal autoantigen 1 (EEA1), transferrin receptor as well as late endosomal markers such as Rab7 and Lamp-1. But *E. coli* $K1^+$ containing vacuoles do not obtain cathepsin D, a

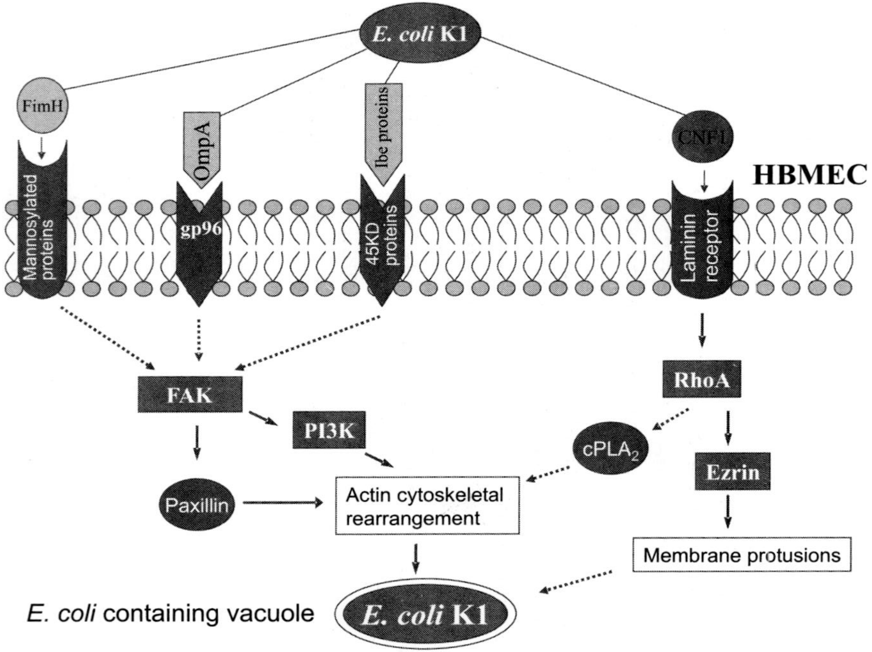


**Fig. 24** Mechanisms of *E. coli* K1 invasion of human brain microvascular endothelial cell. *E. coli* K1 invasion of HBMEC requires actin cytoskeleton rearrangements and activation of FAK, paxillin, PI3K, RhoGTPases and cPLA2. Type 1 fimbriae (FimH), OmpA and IbeA, and their interactions with corresponding receptors in HBMEC contribute to FAK activation. The tyrosine phosphorylation of FAK is upstream of phosphorylation of PI3K and paxillin. This signalling pathway is involved in rearrangments of the actin cytoskeleton required for *E. coli* K1 invasion of HBMEC. CNF1 and its interaction with specific receptor (37LRP/67LR) activate RhoGTPases and subsequently ezrin, and induce the formation of microvilli-like membrane protrusions associated with invading *E. coli* K1. Solid and broken lines represent pathways identified and yet to be identified, respectively (see text for details).

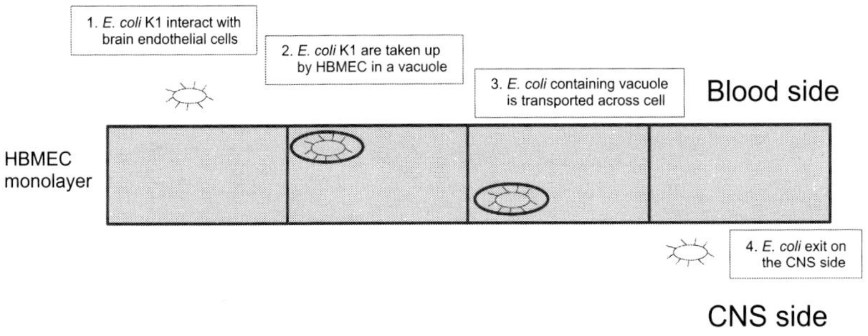


**Fig. 25** Mechanisms of *E. coli* K1 traversal of the blood-brain barrier.

lysosomal enzyme thus avoid bacterial killing by lysosomal enzymes. After crossing the blood-brain barrier as live bacteria, *E. coli* K1 subsequently survives and multiplies in the subarachnoid space, and induces host inflammatory responses in the meninges. Again, *E. coli* $K1^+$ but not $K1^-$ strains can produce positive CSF cultures and cause *E. coli* meningitis indicating that K1 capsule is a critical determinant in *E. coli* meningitis. Although there are several *E. coli* virulence determinants responsible for distinct functions, the involvement of some virulence factors such as OmpA and K1 capsule in more than one step is clear. However, the complete elucidation of the pathogenic mechanisms associated with *E. coli* K1 meningitis is far from complete. The recent availability of *E. coli* K1 genome will undoubtedly help advance our understanding of this fatal infection.

CHAPTER

5

# Protozoa

## 1. INTRODUCTION

The term protozoa is derived from 'proto' meaning 'first' and 'zoa' meaning 'animal'. Protozoa are 'first animals' which generally describes their animal-like nutrition. Protozoa are the largest group of single-celled, microscopic organisms with more than 20,000 species that are found in all aspects of life. Protozoa are widely distributed in various environments from favourable rainforests to sandy beaches to the bottom of oceans to snow-covered mountains. With the availability of improved diagnostic methods, protozoa are being discovered from diverse habitats. However, the abundance and diversity of protozoa in ecosystems is dependent on abiotic factors such as water, temperature, pH, salinity, osmolarity and biotic factors including the availability of food particles. Protozoa include the causative agents of some of the most notorious and deadly diseases. For example, malaria alone causes between one to two million deaths worldwide, annually. Other protozoa play important roles in the food chain maintaining a balanced ecosystem or act as commensal organisms (not harmful) of nearly all humans. Some of the protozoan pathogens have only recently been identified as a major threat to human health. For example, *Cryptosporidium* was originally described in the 19th century, but has recently been associated with serious human infections in AIDS patients. With the increasing number of AIDS patients during the last few decades, many of the protozoan pathogens have become a major problem to human health.

## 2. PROTOZOA: CELLULAR PROPERTIES

Protozoa are the largest single-cell non-photosynthetic animals, which lack cell walls. The study of protozoa, invisible to the naked eye, was initiated with the discovery of the microscope in 1600s by Antonio van Leeuwenhoek (1632 – 1723). The majority of protozoan pathogens are less than 150 μm in size with the smallest one between 1 – 10 μm. The plasma membrane forms the outer boundary of most protozoa and may possess locomotory organelles, such as pseudopodia, flagella and cilia. Although there is diverse structural dissimilarity, the majority of the intracellular organelles are similar to other eukaryotic cells. The DNA is arranged in chromosomes which are contained in one or more nuclei. Protozoa are mostly aerobic and contain mitochondria to generate energy, although many intestinal protozoa are capable of anaerobic growth. The anaerobes lack recognizable mitochondria but may contain hydrogenosomes, mitosomes or glycosomes instead. The cytoplasm particularly in free-living protozoa contains contractile vacuoles. Their function is to regulate osmotic pressure by expelling water. Some protozoa feed by transporting food across the plasma membrane by pinocytosis (engulfing liquids/particles by invagination of the plasma membrane) and/or phagocytosis (engulfing large particles, which may require specific interactions). Some require specialized structures to take up food. For example, ciliates take in food by waving the cilia towards a mouth-like opening known as cytosome. The feeding and/or growing stage is known as trophozoite. However, under harsh conditions, some protozoa transform into a protective cyst form. Cysts can survive lack of food, extremes in pH or temperature, and resist to toxic chemicals or chemotherapeutic compounds. These properties allow some protozoa to find new hosts, thus help in their transmission. Protozoa reproduce asexually by binary fissions (the parent cell mitotically divides into two daughter cells), multiple fission, also known as schizogony (the parent cell divides into several daughter cells), budding, and spore formation, or sexually by conjugation (two cells join, exchange nuclei and produce the progeny by budding or fission). Some protozoa produce gametes (gametocytes, i.e., haploid sex cells), which fuse to form a diploid zygote. Protozoa are among the five major classes of pathogens including intracellular parasites (viruses), prokaryotes, fungi, protozoa and multicellular pathogens. From more than 20,000 species of protozoa, only a handful cause human diseases. However, these few are a major burden on human health and have a severe economic impact. For example, malaria is the fourth leading cause of death, worldwide. To produce disease, protozoa access their hosts via direct transmission through the oral cavity, the respiratory tract, the genito-urinary tract and the skin, or by indirect transmission through insects, rodents as well as by inanimate objects such as towels, contact lenses and surgical instruments. Once the host tissue is invaded, protozoa multiply to establish themselves in the host, and this may be followed by physical damage to the host tissue or depriving it of nutrients, and/or by the induction of an excessive host immune response resulting in disease.

## 3. CLASSIFICATION

Before the availability of molecular tools, taxonomists divided the protozoa into four groups, based on the organisms' mode of locomotion (Fig. 1) as follows.

### 3.1 Phylum Mastigophora

Protozoa placed in this phylum were characterized by the presence of a flagellum, at least during some phases of their life cycles. These organisms are parasitic or free-living in anoxic environments and lack mitochondria and Golgi apparatus. The examples include *Giardia, Trypanosoma, Leishmania, Trichomonas.*

### 3.2 Phylum Ciliophora

The members of this phylum were ciliated during at least one stage of their life cycle, moving typically by the beating of the cilia. They exhibit both asexual and sexual reproduction, i.e., asexual reproduction occurs by budding, binary and multiple fission (schizogony), as well as sexual reproduction by conjugation, autogamy or cytogamy. They are parasitic, commensal or free-living. The examples include *Balantidium, Isotricha, Sonderia.*

### 3.3 Phylum Sarcodina

The majority of protozoa placed in this phylum exhibited movement using characteristic pseudopodia (moving of the protoplasm) into a direction. They are typically, uni-nucleate and possess mitochondria. Reproduction is by asexual fission and they may be parasitic or free-living. The examples include *Entamoeba, Acanthamoeba, Naegleria, Balamuthia, Blastocystis.*

### 3.4 Phylum Apicomplexa

Organisms in this group were all parasitic and characterized by the presence of an apical complex, located at one end of the organisms. They exhibited both sexual and asexual reproduction. The examples include *Plasmodium, Toxoplasma, Babesia, Isospora, Cryptosporidium.*

However this scheme is based on the organisms' locomotion and did not reflect any genetic relatedness. Based on nucleotide sequencing, protozoa are now classified into the following taxa (Fig. 1).

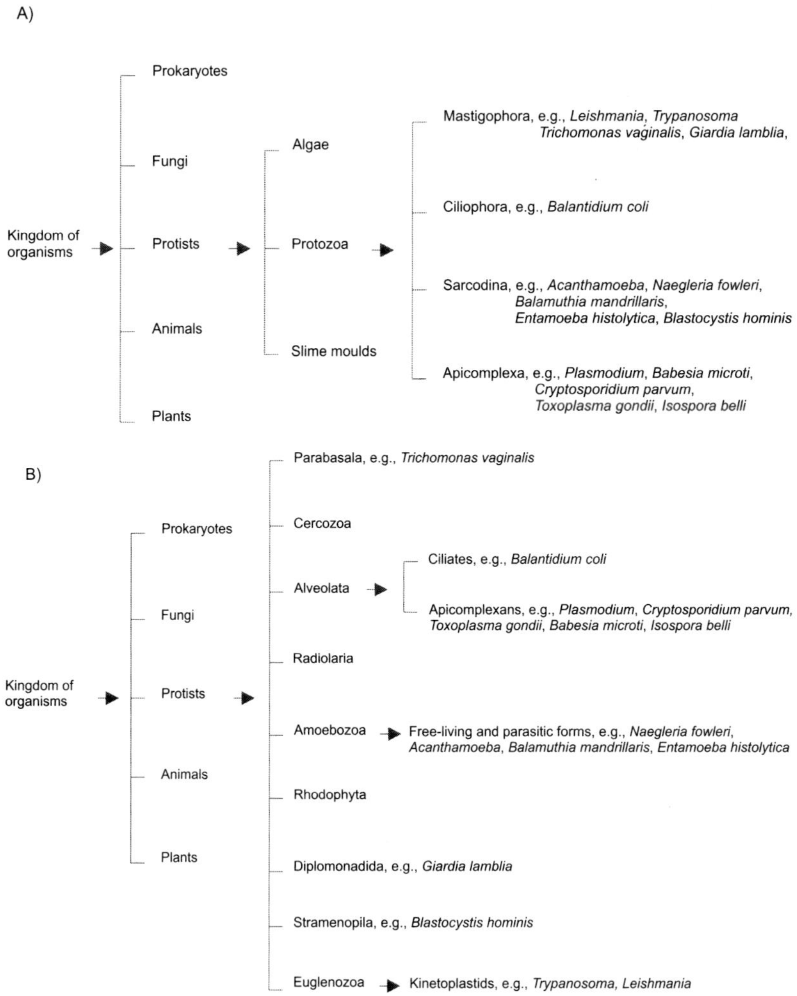


**Fig. 1** **(A)** The old classification scheme of protozoa, based on mode of locomotion. **(B)** The new classification scheme of protists, based on the nucleotide sequencing.

## 3.5 Parabasala

Protozoa placed in this group lack mitochondria, contain a single nucleus and a parabasal body, which is a Golgi body-like structure. The examples include *Trychonympha, Trichomonas*.

## 3.6 Cercozoa

Cercozoa is a group of amoebae with thread-like pseudopodia. These include foraminifera containing a porous shell, composed of calcium carbonate. Pseudopodia extend through holes in the shell. These may be microscopic or several centimeters in diameter. Commonly, foraminifera live of the ocean floors.

## 3.7 Radiolaria

Radiolaria is a group of amoebae that also contain thread-like pseudopodia. Organisms have ornate shells composed of silica and live in marine waters as part of plankton. The pseudopodia of radiolarians radiate from the central body like spokes of a spherical wheel.

## 3.8 Amoebozoa

This group presents the third taxon of amoebae that can be distinguished from the other two taxa by having lobe-shaped pseudopodia and no shells. The examples include *Acanthamoeba, Balamuthia, Naegleria, Entamoeba*. In addition, slime moulds are also included in this group (previously thought to be fungi), based on lobe-shaped pseudopodia, no cell wall ( the cell wall is present in fungi), and feeding. Slime moulds can be further divided into cellular slime moulds such as *Dictyostelium* and acellular slime moulds (also known as Plasmodial slime moulds) that are characterized by filaments of cytoplasm which creep as amoebae and may contain millions of nuclei.

## 3.9 Alveolata

Protozoa in this taxon contain small membrane-bound cavities known as alveoli beneath their cell surface, although the function of these structures remains unclear. This group is further divided into three sub-groups including, (a) ciliates such as *Balantidium coli*, (b) Apicomplexans such as *Plasmodium, Cryptosporidium parvum, Toxoplasma gondii, Babesia microti, Isospora belli*, and (c) dinoflagellates, which are phototrophic such as *Gymnodinium, Gonyaulax*.

## 3.10 Diplomonadida

Protozoa in this taxon lack mitochondria, Golgi bodies and peroxisomes. Organisms have two equal-sized nuclei and multiple flagella, e.g., *Giardia*.

## 3.11 Euglenozoa

This group is further subdivided into two groups; (a) Euglenids, which are photoautotrophic unicellular microbes with chloroplasts-containing pigments (historically thought to be plants). However, they possess flagella, lack a cell wall and are chemoheterotrophic phagocytes (in the dark), e.g., *Euglena*. (b) Kinetoplastids, which have a single large mitochondrion that contains a unique region of mitochondrial DNA, called a kinetoplast. Kinetoplastids live inside animals and some are pathogenic, e.g., *Trypanosoma, Leishmania*.

### 3.12 Stramenopila

This group is a complex assemblage of 'botanical' protists with both heterotrophic and photosynthetic representatives. The evolutionary history of this group is unclear. Generally, the organisms included in this group are slime nets, water moulds and brown algae, and are characterized by possessing flagella. The recent molecular phylogenetic studies revealed that *Blastocystis* belong to this group.

It is important to indicate that no single classification scheme has gained universal support and future studies will almost certainly dictate changes in the above scheme. The representative protozoa pathogens that are covered in this book are indicated in Table 1.

## 4. LOCOMOTION

The motility in protozoa is usually mediated by cilia, flagella, or cellular appendages adapted for propulsion, or amoeboid movement. Other modes of protozoa locomotion involve gliding movements in which no changes in the body shape are observed. The various protozoa have evolved to exhibit distinct movements depending on where they normally live. For example, protozoa with amoeboid movements using pseudopodia are normally present in the environments with abundant organic matter or in flowing water with plant life. Cilia or flagella are used to travel longer distances *per se* that maximizes the possibility of encountering food particles.

### 4.1 Pseudopodia

Pseudopodia are not permanent structures present on the surface of organisms but are formed upon a stimulus. Pseudopodia are observed in amoeboid movement. They are characterized as a flow of cytoplasm in a particular direction. The cytoplasmic membrane temporarily attaches to the substratum and the cytoplasm is drawn into the new attachment protruding as a foot-like structure, hence it is known as pseudopodia. These extensions may exhibit the following distinct phenotypes: 1) broad, round-tipped pseudopodia that is known as the lobopodia, 2) extensively

**Table 1** Representative protozoa pathogens of human importance.

| *Taxon* | | *Human pathogens* | *Disease* | *Source of infection* |
|---|---|---|---|---|
| Parabasala | | *Trichomonas vaginalis* | Urethritis, vaginitis | Contact with vaginal-urethral discharge |
| Alveolata | Ciliates | *Balantidium coli* | Dysentery | Faecal contamination of drinking water |
| | Apicomplexan | *Plasmodium,* | Malaria | Mosquito bite |
| | | *Cryptosporidium,* | Diarrhoea | Humans |
| | | *Toxoplasma,* | Toxoplasmosis | Cats, beef, congenital |
| | | *Babesia,* | Babesiosis | Domestic animals, ticks |
| | | *Isospora* | Coccidiosis | Domestic animals |
| Amoebozoa | | *Acanthamoeba,* | Keratitis, encephalitis | Soil / water |
| | | *Balamuthia,* | Encephalitis | Soil / water |
| | | *Naegleria,* | Encephalitis | Water |
| | | *Entamoeba,* | Dysentery | Faecal contamination of drinking water |
| Diplomonadida | | *Giardia* | Giardial enteritis | Faecal contamination of drinking water |
| Euglenozoa | | *Trypanosoma cruzi,* | Chagas' disease | Triatoma (kissing bug) bite |
| | | *Trypanosoma brucei,* | African trypanosomiasis | Bite of tsetse fly |
| | | *Leishmania* | Leishmaniasis | Bite of sand fly |
| Stramenopila | | *Blastocystis hominis* | Diarrhoea | Contaminated food / water |

branched forming a net-like structure which is known as the rhizopodia, 3) sharp, pointed projections that are the filopodia, and 4) axopodia, i.e., similar to filopodia but contain slender filaments. Pseudopodia play important roles both in locomotion as well as in food uptake. The examples include *Acanthamoeba* spp. *and Entamoeba histolytica.*

## 4.2 Cilia and Flagella

In contrast to pseudopodia, flagella and cilia are permanent microtubular organelles that are anchored within the plasma membrane of certain protozoa, and project from the cell surface. Flagella are long slender structures (50 – 200 μm) usually one to a few on a cell, with whip-like movements starting at the tip or the base of the cell. This results in a forward, backward or spiral movements. Cilia are similar to flagella but smaller in size, i.e., 5 – 20 μm and usually more numerous. Cilia move with a back-and-forth stroke. Their thickness is approximately 0.2 μm. Both cilia and flagella exhibit a bending motion resulting in fluid propulsion and cellular movements. However, due to the extended length of the flagellum, bends are propagated along the flagellum pushing the surrounding water symmetrically on both sides of the cell. In contrast, with the shorter length and large numbers, cilia beat in co-ordination with one another.

### 4.3 Gliding Movements

Other forms of locomotion in protozoa are gliding. For example, several flagellates glide over surfaces. In such cases, flagella do not exhibit a whip-like movement but make contact with the surface and slide over it with the aid of microtubules. Other examples include sporozoites stage during the life cycle of *Plasmodium* spp.

### 4.4 Locomotory Proteins

The major proteins involved in locomotion are (i) microtubules that are cylindrical fibrils formed of tubulin molecules around 25 nm in diameter, and (2) microfilaments, also known as actin filaments which are composed of actin molecules and are about 7 nm thick. The polymerization and depolymerization of the tubulin molecules in microtubules and their ability to interact with dynein adenosine triphosphatase proteins cause movement in protozoa. In contrast, actin is polymerized and forms microfilaments. These microfilaments are pushed along by interacting with myosin ATPase molecules. Other proteins important in protozoa locomotion are myonemes that are centrin, which form filaments of about 10 μm thickness. Depending on calcium concentration, centrin exist in two states: Filaments shorten in the presence of calcium (centrin binds to calcium) and extend when the calcium is withdrawn. These contractions result in cellular movements, e.g., *Vorticella, Stentor.*

## 5. FEEDING

Protozoa consists of both photosynthetic organisms (autotrophs which synthesize their own food), non-photosynthetic organisms (heterotrophs that obtain food such as organic molecules or particulate matters from the environment) and organisms which use both modes and are known as mixotrophs. Photosynthetic protozoa usually occur among the flagellates. Some protozoa, i.e., certain ciliates, temporarily acquire chloroplasts from ingested algae and use them to synthesize food. In other cases, protozoa maintain a symbiotic relationship with another organism for their nutrition needs. The partner in the symbiotic relationship may involve a virus, a prokaryote, a protist or a multicellular eukaryote. The resultant outcome of this relationship may be commensalisms, i.e., a condition in which one organism benefits. Usually the smaller organism feeds on the food that is unusable to or unwanted by the host. For example, *Entamoeba coli* feeds on bacteria in the large intestine of humans. Other relationships may be mutualism, in which, both partners are nutritionally dependent on each other. For example, termites are unable to digest cellulose and thus cannot live without the cellulolytic bacteria flagellates in their digestive tract, which obtain nutritional benefits in return. And lastly parasitism, a condition in which one organism lives at the expense of the other by obtaining essential nutrients. It is widely accepted that the 'true parasites' do not intend to kill their host and maintains a balanced relation with the host. However, the overwhelming hosts' immune response and/or burden of parasite could result in the disease and even in the host's death. Bacteria form the major food source for the majority of protozoa including ciliates, flagellates and amoebae. The mode of nutrient uptake is dependent on their nature. For example, particles are taken up by phagocytosis. Phagocytosis is a process in which particles are bound to the cell surface or captured by pseudopodia, followed by invagination of the plasma membrane, thus engulfing the food particle in an intracellular vacuole. These vacuoles are fused with lysosomes containing digestive enzymes which degrade the food particles. In contrast, fluid or dissolved nutrients are taken up by pinocytosis or transported across the plasma membrane by diffusion or active transport. These processes are facilitated by cytoskeletal rearrangements involving microtubules and actin filaments. Some protozoa, e.g., *Amoeba proteus* or *Balamuthia* are capable of attacking and ingesting other protozoa.

## 5.1 Metabolism

The non-photosynthetic (heterotrophic) protozoa engulf smaller organisms or organic matter and digest them to generate energy to perform their routine functions. Alternatively, protozoa use waste organic particles produced by the metabolism of other organisms. Protozoa use inorganic solutes including potassium, chloride, essential metals, nitrates, ammonium and amino acids, inorganic phosphates as well as organic phosphate compounds, ethanol, other short chain alcohols, and organic acids such as acetate, pyruvate and lactate as carbon sources for their cellular metabolism. However, the photosynthetic (autotrophic) protozoa can metabolize organic particles or use inorganic solutes to synthesize their food. Many catabolic (pathways involved in degradation) and anabolic (synthetic) pathways are similar to those found in other eukaryotes.

### *5.1.1 Glycolysis*

The majority of protozoa metabolize carbohydrates as energy sources and exhibit minimal catabolism of fatty acids and amino acids. They maintain intracellular stores of carbohydrates in the form of glycogen (or trehalose). Usually the bloodstream parasites such as *Plasmodium* and *Trypanosoma* do not store polysaccharides as they have ready access to glucose. Glucose or glycogen is broken down by glycolysis to give pyruvate. In *Trypanosoma* most of the glycolytic pathway is sequestered in an organelle, the glycosome, and its link with a mitochondrion is shown in Fig. 2. This adaptation allows glycolysis to function optimally. In some protozoa, which are unable to utilize glucose, organic acids or short chain alcohol are used as alternative carbon sources to glucose.

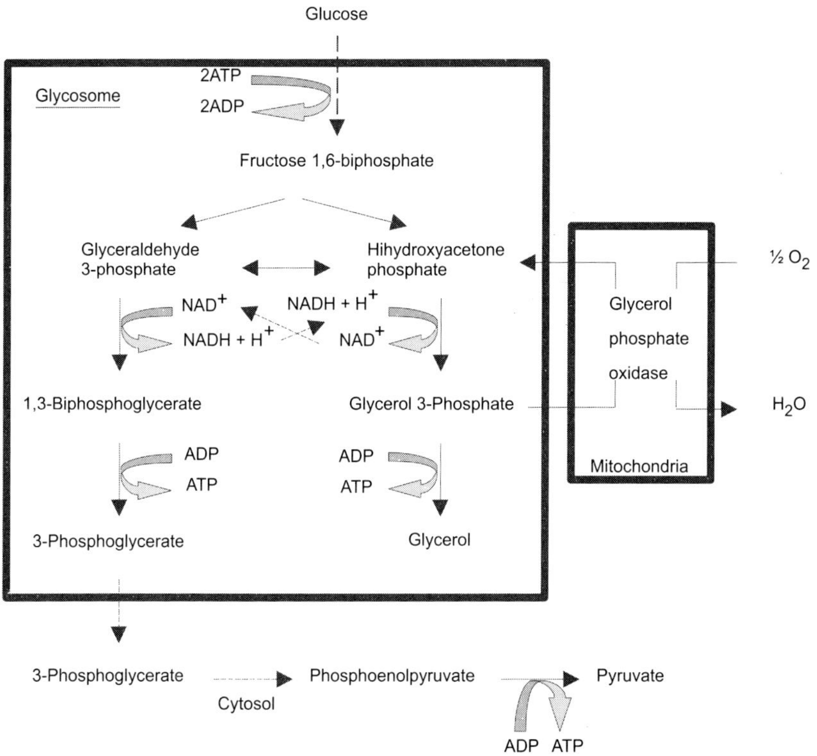


**Fig. 2** Aerobic glycolysis in bloodstream forms of *Trypanosoma*, showing compartmentalization of much of the pathway in the glycosome. Dotted lines indicate several reaction steps.

### *5.1.2 Mitochondrial metabolism*

In aerobic protozoa, mitochondria have electron transport system exhibiting mammalian-like respiration sensitive to cyanide and having a cytochrome $aa_3$ as terminal oxidase. The glycolytic products enter the mitochondria and generate adenosine triphosphate (ATP) via the oxygen dependent cycle similar to the mammalian electron transport system. Usually protozoa do not fully degrade end products of metabolism to carbon dioxide and water as observed in the mammalian system. These differences may have potential implications in the identification of novel pathways to inhibit parasite energy metabolism, which may prove fatal for parasites and offer targets for therapy.

### *5.1.3 Hydrogenosomal metabolism*

Protozoa that inhabit oxygen poor environments (anaerobes) such as *Trichomonas* do not contain mitochondria but possess hydrogenosome organelle. In such organisms, pyruvate or malate formed by glycolytic pathways enter the hydrogenosomes and are converted by fermentation reactions into a range of incompletely oxidized compounds such as acetate with the production of ATP. In addition, during this process, molecular hydrogen is generated by the action of an anaerobic electron-transport pathway (Fig. 3).

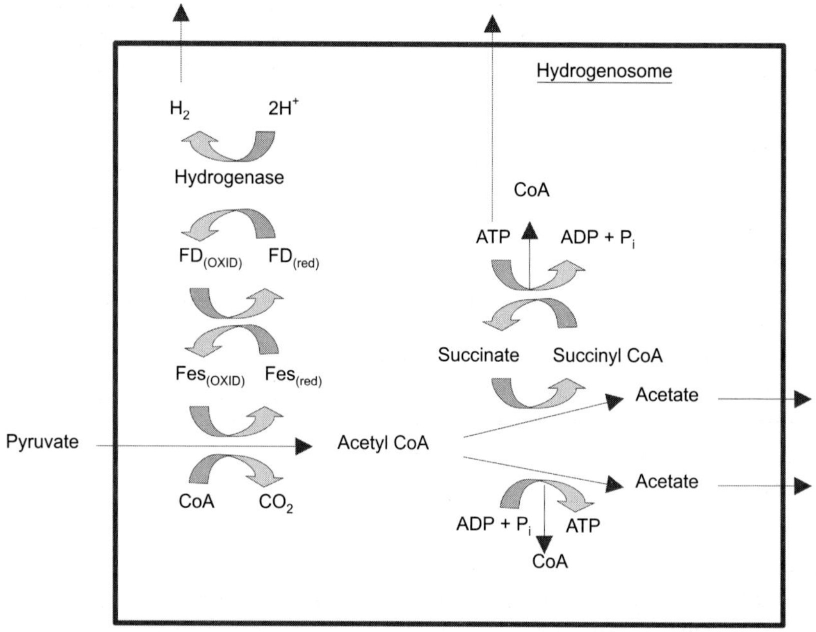


**Fig. 3** Hydrogenosomal metabolism in the Trichomonads. Fes = Iron-sulphur proteins; Fd = ferridoxins; oxid = oxidized state; red = reduced state.

### 5.1.4 *Nitrogen metabolism*

Free-living protozoa can utilize inorganic as well as organic sources of nitrogen when available. Of the inorganic sources, ammonium is preferentially used. However, many species have nitrate reductase (converts nitrate to nitrite) and nitrite reductase (converts nitrite to ammonium) suggesting that these organisms can use many available sources of nitrogen. Most parasitic protozoa and those involved in symbiotic relationships will exclusively use amino acids as sources of nitrogen. Parasites secrete protease enzymes that degrade proteins releasing small peptides and amino acids. Amino acids may also be end products of metabolism and may be excreted; storage and excretion of amino acids may play a role in osmoregulation in protozoan lacking contractile vacuoles.

Overall, protozoa employ diverse mechanisms for adaptation of their energy metabolism. A complete understanding of the biochemical pathways associated with protozoa energy metabolism is crucial for the rationale development of anti-parasitic drugs.

## 6. REPRODUCTION

To ensure survival, a species must reproduce in large numbers when conditions are favourable. The reproduction in protozoa can be sexual, i.e., formation and fusion of gametes producing new offsprings or asexual, i.e., mitotic division of a parent cell into two or more identical offsprings. Many protozoa only use asexual, while others use both during their life cycles.

## 6.1 Asexual Reproduction

Asexual reproduction is the primary mode of reproduction in many protozoa. Asexual reproduction occurs by binary fission, multiple fission (schizogony), and budding. The process of cell division is divided into an S phase (DNA synthesis phase) and an M phase (division of nucleus that involves DNA condensation and organized distribution of chromosomes). Both phases are separated by gap phases, i.e., G1 and G2. In addition, cytoplasmic organelles are duplicated, followed by cytokinesis. The resulting daughter cells are identical to their parent cell and are produced in sufficient numbers for successful transmission. Asexual reproduction occurs in majority of amoebae, ciliates and flagellates.

### 6.1.1 *Binary fission*

In this form, a single cell divides into two daughter cells and it is the most common type of asexual reproduction lasting from 6 – 24 h. This results in very large numbers of identical parasite populations within days. The DNA and cytoplasmic organelles are duplicated followed by nuclear division and cytokinesis, and finally a constriction ring bisects the cell producing two daughter cells. The examples include *Toxoplasma* and *Acanthamoeba.*

### 6.1.2 Multiple fission (Schizogony)

In this form of reproduction, mitotic nuclear division occurs several times. This results in several nuclei within the cytoplasm. Each nucleus together with a layer of cytoplasm gives rise to independent daughter cells that are released. In parasitic protozoa, it results in the rapid production of large protozoan populations which overwhelms the host immune system. This occurs in various protozoa including *Pelomyxa palustris, Volvox, Gonium.*

### 6.1.3 Budding

In this form of reproduction, the nuclei divide and the daughter nuclei migrate into a cytoplasmic bud. This is followed by cytoplasmic fission and release of the cell, which develops into a mature reproductive organism. The resulting daughter cells may differ from the parent cells. Such reproductive schemes are limited to several protozoa such as *Trichophrya, Ephelota.*

## 6.2 Sexual Reproduction

Many parasitic protozoa have complex life cycles and reproduce both sexually and asexually. Sexual reproduction involves formation of haploid gametes by meiosis that fuse to form a diploid zygote generating new organisms.

### 6.2.1 Gamete formation

Gametes with haploid chromosomes are formed in the vegetative stage, which fuse with one another to produce diploid zygotes. The zygote undergoes meiosis and the numbers of chromosomes are halved followed by asexual reproduction to produce large populations of the organisms. Gametes formation directly from the original population followed by their fusion into a zygote and is called 'hologamy'. Examples include *Chlamydomonas, Dunaliella* and *Polytoma.* If gametes are identical (at least morphologically but there may be minor genetic or physiological differences), the term 'isogamy' is used. These include *Chlorogonium* as well as many foraminiferan sarcodinids produce isogametes and some sporozoa. When there are clear differences between male and female gametes, typically known as micro- and macrogametes (size differences, presence of flagella, physiological or biochemical properties), the gamete formation and their fusion into a zygote is called 'anisogamy'. This is a common form of gamete formation in protozoa. Examples include *Plasmodium* spp. The factors that induce gamete formation are not clearly known but may involve environmental conditions such as salinity, pH, temperatures, nutrients, etc.

If both gametes arise from the same clone, the species is called monoecious but if they arise from different clones, the species is called diecious. However, if self-fertilization occurs, the species is known as hermaphrodite. If the new population of an organism develops from unfused gamete without fertilization, it is called parthenogenesis. The examples include genus *Volvox*, *Eucoccidium dinophili*.

#### 6.2.2 *Gametic nulei: Conjugation*

Some ciliated protozoa are unable to produce gametes. Instead they possess dual nuclei and during their sexual reproduction, the nuclei from two organisms (instead of gametes) fuse together yielding a zygotic nucleus. This is followed by asexual fissions producing large populations. Again, diverse factors are responsible for conjugation including temperatures, salinity, pH, etc. Examples include *Paramecium* and *Tetrahymena*.

## 7. LIFE CYCLE

The protozoa that are major human pathogens can be broadly subdivided into three main categories, i.e., those which are blood-borne that are usually transmitted via insects (e.g., *Plasmodium*); those which are transmitted via contaminated food/water (e.g., *Giardia*) and those which are sexually transmitted (e.g., *Trichomonas*). The life cycles of blood-borne protozoa generally involves humans and insects. The life cycle in insects may be necessary for parasite development or merely acts as vectors to transmit to a new host. In contrast, other protozoa pathogens are acquired by the new host through exposure to contaminated water and/or food or through sexual intercourse.

### 7.1 *Plasmodium* spp.

Although the precise life cycle of *Plasmodium* varies between species, it can be divided broadly into two hosts as follows.

#### 7.1.1 *Life cycle in the vertebrate host*

When an infected mosquito takes a blood meal, it injects saliva containing sporozoites (approx. 10 – 15 μm long and 1 μm in diameter). The sporozoites penetrate hepatocytes (liver cells) and undergo asexual reproduction, a process known as the pre-erythrocytic (PE) cycle or exoerythrocytic cycle (EE). Within the hepatic cell, the parasite transforms into a trophozoite stage that feeds on the host cell cytoplasm. Within a few days, the trophozoite matures and produces a daughter nuclei and at this stage this is called a schizont. A single schizont undergoes cytokinesis and produces many daughter cells called merozoites (2.5 μm long and 1.5 μm in diameter). The merozoites are released and infect new hepatocytes or enter the erythrocytic cycle. Upon entry into an erythrocyte, a merozoite again transforms into a trophozoite and feeds on the host cell cytoplasm, forming a large food vacuole giving a characteristic ring appearance. The trophozoite becomes a schizont again and produces many merozoites which infect new erythrocytes. After several generations, some merozoites infect erythrocytes and become macrogametocytes and microgametocytes. These are taken by the mosquito, where the remaining life cycle continues (Fig. 4).

### *7.1.2 Life cycle in the invertebrate host*

The gametocytes are taken up by the mosquito during their blood meal. If gametocytes are taken up by a susceptible mosquito (*Anophles* spp. in the case of human *Plasmodium*), gametocytes develop into gametes, i.e., micro- and macro-gametes. The nucleus of microgametocyte divides to produce 6 to 8 nuclei and exflagellates. The microgamete fuses with a macrogamete to form a diploid zygote. The zygote elongates to become a motile ookinete that is 10 – 15 μm. The ookinete penetrates the gut wall of the mosquito and develops into an oocyst. The oocyst undergoes meiosis and produce sporozoites, which penetrate salivary glands and are injected into new a host at the next blood meal, thus completing the cycle (Fig. 4).

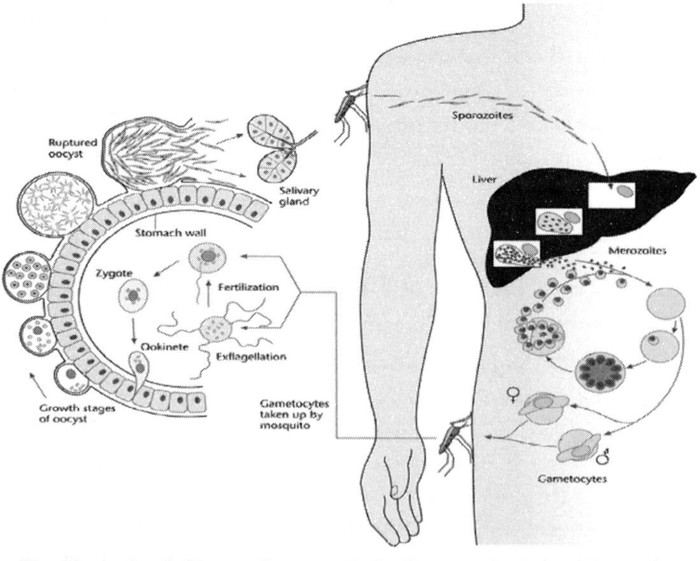


**Fig. 4** The life cycle of *Plasmodium* spp. indicating vertebrate and invertebrate hosts. *Colour image of this figure appears in the Colour Plate Section at the end of the Book.*

## 7.2 *Trypanosoma brucei*

As for *Plasmodium*, it can be divided broadly into two hosts which is as follows.

### 7.2.1 Life cycle in the vertebrate host

When an infected tsetse fly (*Glossina* spp.) takes a blood meal, it injects trypomastigotes. The trypomastigotes then travel through the lymphatic and circulatory systems to other sites, where they reproduce by binary fission. Eventually, some trypomastigotes enter the central nervous system and the cerebrospinal fluid, while other continue to circulate to be picked up by the tsetse fly, where the remaining life cycle continues (Fig. 5).

### 7.2.2 Life cycle in the invertebrate host

The trypomastigotes are taken up by tsetse fly during their blood meal. In the mid-gut of the fly, trypomastigotes multiply by binary fission, producing immature epimastigotes that migrate to the salivary glands. Here, they mature and become trypomastigotes and are injected into a new host at the next blood meal, thus completing the cycle (Fig. 5).

## 7.3 *Trypanosoma cruzi*

### 7.3.1 Life cycle in the vertebrate host

Trypomastigotes are shed in the faeces of the kissing bug (*Triatoma* spp.), while the bug feeds on the mammalian host. When the host scratches the itchy wound, trypomastigotes enter the wound and then travel throughout the body and penetrate certain cells, especially macrophages and the heart muscle cells, where they transform into non-flagellated amastigotes. The amastigotes divide by binary fission and eventually rupture the host cells. The released amastigotes either infect new cells or transform into trypomastigotes that circulate in the bloodstream to be picked up by the kissing bug, where the remaining life cycle continues (Fig. 6).

### 7.3.2 Life cycle in the invertebrate host

The circulating trypomastigotes are taken up by the kissing bug during their blood meal. In the mid-gut of the bug, trypomastigotes multiply by binary fission, producing immature epimastigotes that migrate to the hindgut of the kissing bug. Here, they mature and become trypomastigotes and are injected into a new host at the next blood meal, thus completing the cycle (Fig. 6).

## 7.4 *Leishmania*

### 7.4.1 Life cycle in the vertebrate host

When an infected sandfly (*Phlebotomus* spp. or *Lutzomyia* spp.) takes a blood meal, it injects promastigotes. Macrophages near the bite site

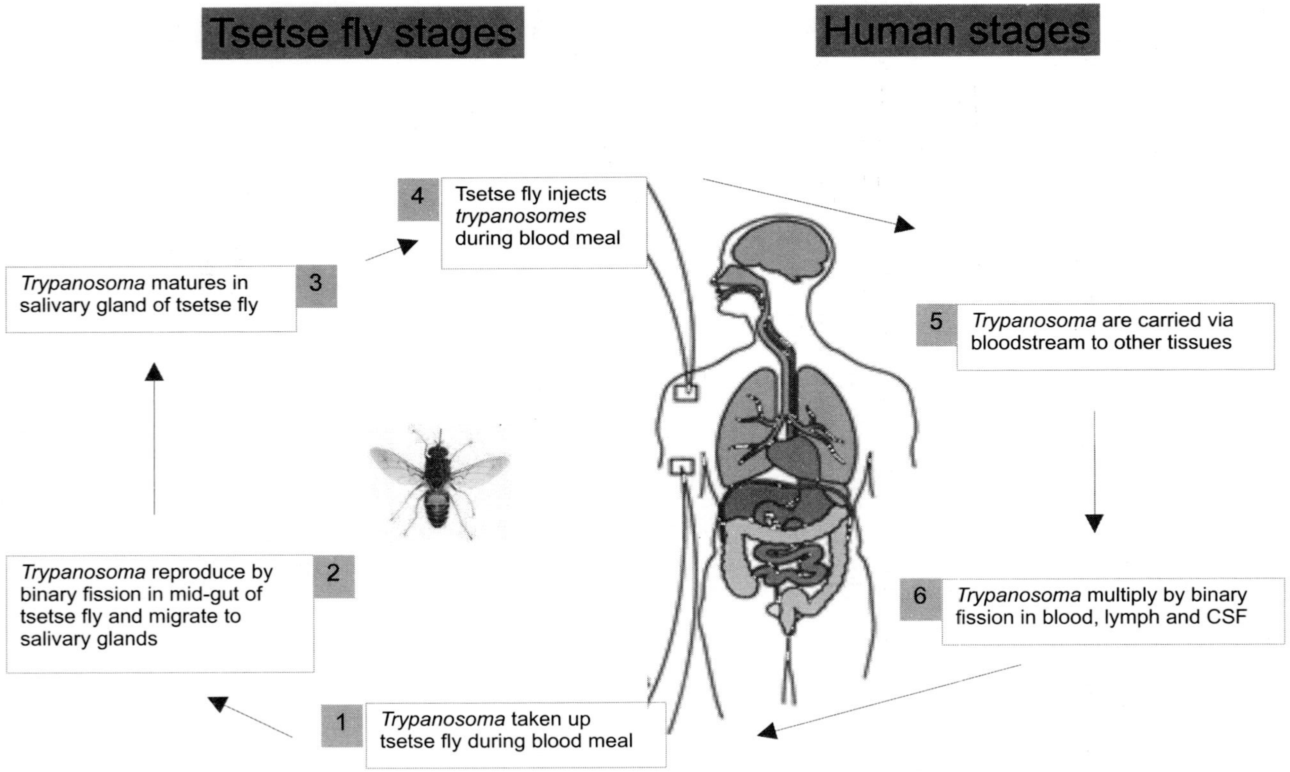


**Fig. 5** The life cycle of *Trypanosoma brucei* indicating vertebrate and invertebrate hosts.
*Colour image of this figure appears in the Colour Plate Section at the end of the Book.*

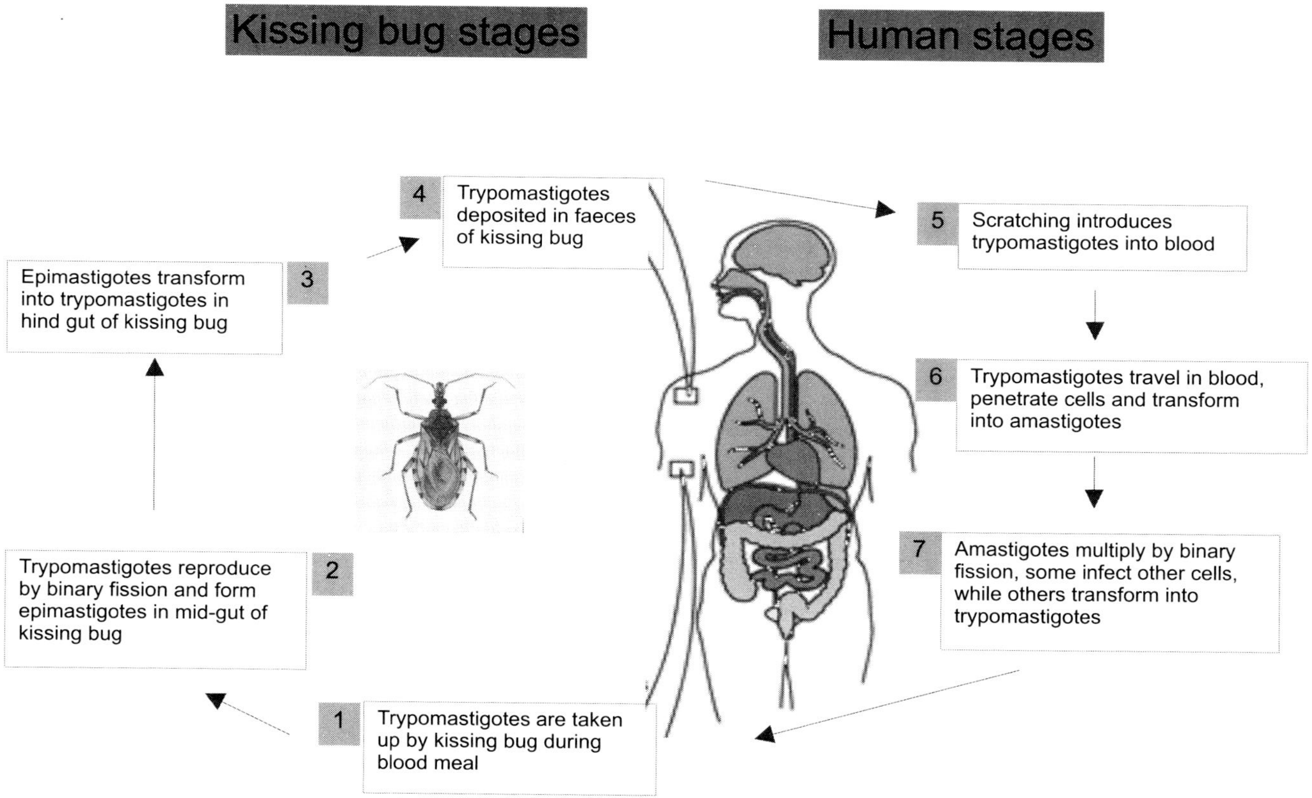


**Fig. 6** The life cycle of *Trypanosoma cruzi* indicating vertebrate and invertebrate hosts.
*Colour image of this figure appears in the Colour Plate Section at the end of the Book.*

phagocytoze the promastigotes, which then transform into amastigotes. The amastigotes are reproduced by binary fission until the macrophage ruptures. The released amastigotes infect other macrophages as well as circulate in the bloodstream to be picked up by the sandfly, where the remaining life cycle continues (Fig. 7).

#### *7.4.2 Life cycle in the invertebrate host*

During the blood meal, the sandfly takes up phagocytes containing amastigotes. In the mid-gut of the fly, amastigotes are released from the phagocytes and transform into promastigotes. Promastigotes rapidly divide by binary fission, filling the fly's digestive tract and migrate to the proboscis, and are injected into a new host at the next blood meal, thus completing the cycle (Fig. 7).

## 8. PROTOZOA AS HUMAN PATHOGENS

Protozoa are responsible for the most notorious and deadly diseases killing millions of animals and humans. Among the protozoa, the following are the important human pathogens.

### 8.1 Flagellates

#### *8.1.1 Organism: Trichomonas vaginalis*

**Biology:** Trophozoites range from 10 – 25 μm in length, contain a single nucleus, phagocytose bacteria and leukocytes as a food source and reproduce by binary fission but do not form cysts.

**Disease:** Infection of the urogenital systems. Usually asymptomatic but when acquired these parasites attach to the epithelial cells and can produce severe inflammation or swelling of the sexual organs.

**Diagnosis:** Diagnosis is made by demonstrating parasites by microscopy.

**Transmission:** Sexual intercourse with infected individuals.

**Treatment:** Oral application of metronidazole or tinidazole is effective.

**Occurrence:** Worldwide.

#### *8.1.2 Organism: Giardia lamblia*

**Biology:** Trophozoites are approximately 12 – 15 μm long, contain two nuclei, feed on nutrients obtained from the intestinal fluid and reproduce by binary fission. They form cysts for transmission.

**Disease:** It causes diarrhoea. Parasites stick to the intestinal epithelium but do not lyse host cells. Diarrhoea is observed within a few days to

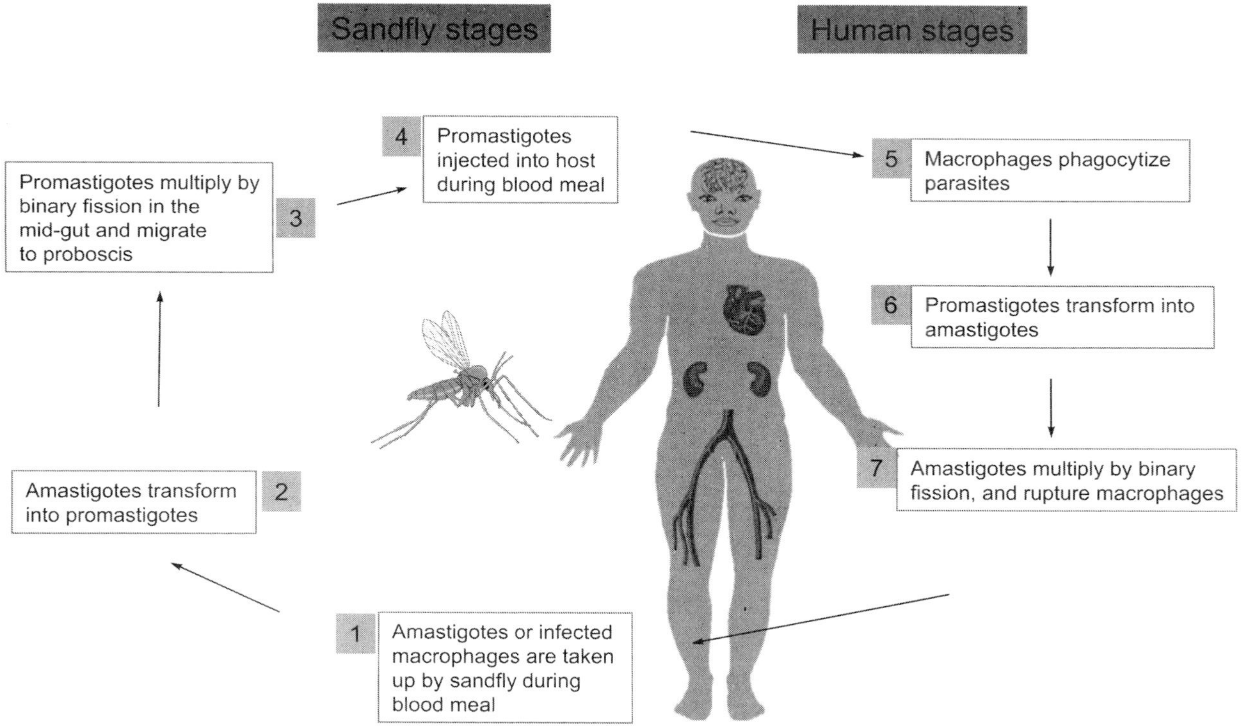


**Fig. 7** The life cycle of *Leishmania* spp. indicating vertebrate and invertebrate hosts.
*Colour image of this figure appears in the Colour Plate Section at the end of the Book.*

weeks but without blood. Other symptoms include intestinal pain, dehydration and weight loss but not fatal.

**Diagnosis:** Diagnosis is made by demonstrating parasites by microscopy or serology-based assays.

**Transmission:** Oral uptake of *Giardia* cysts from human faeces in contaminated food or drinking water.

**Treatment:** Oral application of metronidazole is effective against the trophozoite.

**Occurrence:** Worldwide.

### *8.1.3 Organism: Trypanosoma brucei gambiense / Trypanosoma brucei rhodesiense*

**Biology:** Trophozoites are approximately 18 – 29 µm long, contain a single nucleus, feed on nutrients obtained from the host and are reproduced by binary fission but do not form cysts.

**Disease:** Sleeping sickness or African Trypanosomiasis. Symptoms are observed within weeks and include fever, headache, swollen lymph nodes, weight loss, and heart involvement (usually with *T. b. rhodesiense*). It is important to note that *T. b. rhodesiense* causes an acute disease and the host often dies before the disease can develop fully. In contrast, *T. b. gambiense* produces a chronic disease and the parasite invades the CNS and produces typical symptoms such as tremors, sleepiness, paralysis and finally death within months. Of interest, *T. b. brucei* causes fatal infections in non-human mammals, such as horses, sheep, etc.

**Diagnosis:** Based on symptoms and demonstration of parasites by microscopy.

**Transmission:** Disease transmission occurs via insect bites, i.e., tsetse fly (vector) where parasites are present in the salivary glands.

**Treatment:** Use of suramin, pentamidine, Berenil and difluoromethylornithine early in the disease may have beneficial effects and melarsoprol at later stages.

**Occurrence:** Africa.

### *8.1.4 Organism: Trypanosoma cruzi*

**Biology:** Trophozoites are approximately 20 µm long, contain a single nucleus, feed on nutrients obtained from the host and are reproduced by binary fission but do not form cysts.

**Disease:** Chagas' disease or American Trypanosomiasis. Following invasion, parasites infect susceptible tissues in particular the heart. Parasites penetrate the myocardial fibres, multiplying for several days and producing a cavity in the invaded tissue and escape into the bloodstream and invade other susceptible tissues including the liver, spleen, muscles, intestinal mucosa resulting in organ failure and finally death within a few months to years.

**Diagnosis:** Based on symptoms and demonstration of parasites by microscopy.

**Transmission:** The disease transmits via insects, i.e., triatomid bugs (vector). While feeding, these bugs defecate on the host, and parasites in the faecal material gain entry into the human body through the bite wound.

**Treatment:** No effective treatment but use of nifurtimox, ketoconazole and benznidazole may be effective.

**Occurrence:** South and Central America.

### *8.1.5 Organism: Leishmania tropica / Leishmania major*

**Biology:** Trophozoites are approximately 2 – 5 μm long, contain a single nucleus, feed on nutrients obtained from the host and reproduce by binary fission but do not form cysts.

**Disease:** Cutaneous leishmaniasis. Following invasion, parasites multiply within reticuloendothelial and lymphoid cells. Symptoms include cutaneous lesions within days to months at the site of insect bite.

**Diagnosis:** Microscopic demonstration of parasites.

**Transmission:** The disease transmits via insects, i.e., sandfly. While feeding, these bugs excrete on the host, which contain parasites. These parasites gain entry into the human body through the site of the bite.

**Treatment:** Injections of pentostam and glucantime provide effective treatment. In addition, sodium stibogluconate may be useful.

**Occurrence:** Africa, Middle East, Asia.

### *8.1.6 Organism: Leishmania donovani*

**Biology:** Trophozoites are approximately 2 – 5 μm long, contain a single nucleus, feed on nutrients obtained from the host and are reproduced by binary fission but do not form cysts.

**Disease:** Kala-azar or visceral leishmaniasis. Following invasion, parasites are taken up by the macrophages and multiply, eventually killing the macrophages and severely affecting the host defences. Symptoms appear within a few weeks to months and include fever, malaise, anemia, an enlarged liver and spleen and finally death. Of interest, *L. donovani* taken up by neutrophils are killed but this does not have a major effect on the outcome of the disease.

**Diagnosis:** Microscopic demonstration of parasites.

**Transmission:** The disease transmits via insects, i.e., sandfly.

**Treatment:** Injections of pentostam and glucantime provide effective treatment. In addition, sodium stibogluconate may be useful.

**Occurrence:** The Americas, Africa, Asia, the Mediterranean.

## 8.2 Amoebae

### *8.2.1 Organism: Entamoeba histolytica*

**Biology:** Trophozoites are about 20 – 30 µm long, contain a single nucleus, feed on nutrients obtained from the host, reproduce by binary fission and form cysts for transmission.

**Disease:** Diarrhoea or amoebic dysentery and liver abscesses. Symptoms include intestinal lesions which develop in the gut and with an increasing number of parasites, the mucosal destruction becomes extensive with abdominal pain, diarrhoea (bloody), cramps, vomiting, malaise, weight loss with extensive scarring of the intestinal wall. Death can occur with gut perforation, exhaustion and liver abscesses.

**Diagnosis:** Diagnosis is made by direct demonstration of parasites in the stool samples or using serology-based assays.

**Transmission:** Oral uptake of *Entamoeba* cysts from human faeces with contaminated food or drinking water.

**Treatment:** Treatment includes oral application of metronidazole or diiodohydroxyquin and iodoquinol.

**Occurrence:** Worldwide ( more prominent in warm countries).

### *8.2.2 Organism: Acanthamoeba* spp. *causing blinding keratitis*

**Biology:** Trophozoites are about 15 – 35 µm long, contain a single nucleus, feed on bacteria, are reproduced by binary fission and form cysts under harsh conditions as well as for transmission.

**Disease:** Blinding keratitis, usually associated with contact lens use. Initially irritation of the eye, inflammation, photophobia, epithelial defects, excruciating pain, and can result in blindness.

**Diagnosis:** Microscopic demonstration of parasites and culturing.

**Transmission:** Washing contact lenses with tap water or home-made saline is a common mode of transmission. Swimming/sleeping with contact lenses or extended wear of contact lenses are additional predisposing factors.

**Treatment:** Successful treatment requires early diagnosis with aggressive application of polyhexamethylene biguanide (PHMB) or chlorhexidine digluconate (CHX) together with propamidine isethionate, also known as Brolene.

**Occurrence:** Worldwide.

### *8.2.3 Organism: Acanthamoeba* spp. *causing fatal encephalitis*

**Biology:** Trophozoites are about 15 – 35 μm long, contain a single nucleus, feed on bacteria, are reproduced by binary fission and form cysts under harsh conditions as well as for transmission.

**Disease:** Granulomatous encephalitis involving the CNS (limited to immunocompromised patients). Infection initiates with amoebae entry into the bloodstream via the lower respiratory tract or directly through skin lesions followed by their crossing of the blood-brain barrier into the CNS to produce the disease. Alternative routes of entry involve olfactory neuroepithelium. The clinical syndromes include meningitis-like symptoms such as headache, fever, behavioural changes, lethargy, a stiff neck, vomiting, nausea, increased intracranial pressure, seizures and finally leading to death.

**Diagnosis:** Microscopic demonstration of parasites and serology-based assays.

**Transmission:** Limited to immunocompromised patients. Their exposure (especially in the presence of skin lesions) to standing water, soil, mud, swimming in contaminated water, i.e., lakes, untreated pools may attribute to this fatal infection.

**Treatment:** No recommended treatment, usually fatal, current therapeutic agents include a combination of ketoconazole, fluconazole, sulfadiazine, pentamidine, amphotericin B, azithromycin, or itraconazole. Alkylphosphocholine compounds, such as hexadecylphosphocholine have shown promising results.

**Occurrence:** Worldwide.

### *8.2.4 Organism: Balamuthia mandrillaris*

**Biology:** Trophozoites are about 20 – 45 μm long, contain a single nucleus, feed on eukaryotic cells, are reproduced by binary fission and form cysts under harsh conditions as well as for transmission.

**Disease:** Granulomatous encephalitis involving the CNS. Unlike *Acanthamoeba*, it can cause infections in relatively immunocompetent individuals. Infection initiates when the amoebae enter the bloodstream via the lower respiratory tract or directly through skin lesion followed by their crossing of the blood-brain barrier into the CNS to produce the disease. Alternative routes of entry are via the olfactory neuroepithelium. Symptoms include headache, fever, behavioural changes, nausea, a stiff neck, photophobia, seizures and finally leading to death within a few weeks to months.

**Diagnosis:** Microscopic demonstration of parasites and serology-based assays.

**Transmission:** Exposure (especially in the presence of skin lesions) to soil, mud, during gardening, as well as swimming in contaminated water may attribute to this fatal infection.

**Treatment:** No recommended treatment, usually fatal, current therapeutic agents include a combination of ketoconazole, fluconazole, sulfadiazine, pentamidine, amphotericin B, azithromycin, or itraconazole.

**Occurrence:** Worldwide.

### *8.2.5 Organism: Naegleria fowleri*

**Biology:** Trophozoites are about 15 – 35 μm long, contain a single nucleus, feed on bacteria, are reproduced by binary fission and form cysts under harsh conditions as well as for transmission.

**Disease:** Primary amoebic meningoencephalitis involving the CNS. Infection initiates when amoebae enter the nasal passage (during swimming/diving into the water). Amoebae migrate along the olfactory nerves, through the cribiform plate into the cranium. Death occurs within days.

**Diagnosis:** Microscopic demonstration of parasites and serology-based assays.

**Transmission:** Exposure to the contaminated water, i.e., swimming/diving.

**Treatment:** No recommended treatment, usually fatal, current therapeutic agents include amphotericin B or qinghaosu.

**Occurrence:** Worldwide.

### *8.2.6 Organism: Blastocystis hominis*

*Blastocystis hominis* (previously thought to be protozoa but more closely related to other protists within the kingdom Chromista, based on rRNA sequences).

**Biology:** Trophozoites vary greatly in size from 5 – 40 µm, contain a single nucleus, are reproduced by asexual reproduction and form cysts under harsh conditions as well as for transmission.

**Disease:** Blastocystosis (especially in patients with weak immune system). Usually asymptomatic, however in a minority of individuals it can cause diarrhoea, loose stools, anal itching, abdominal pain and weight loss but it has not been proven whether symptoms come from *Blastocystis* spp. or as a result of other (bacterial/viral) infections.

**Diagnosis:** Diagnosis is made by direct demonstration of parasites in the stool samples.

**Transmission:** Faecal-oral route through contaminated food/water. Parasite divides by binary fission in the GI tract.

**Treatment:** Treatment involves oral application of metronidazole for prolonged periods of time.

**Occurrence:** Worldwide.

## 8.3 Sporozoa Apicomplexa

### *8.3.1 Organism: Toxoplasma gondii*

**Biology:** Trophozoites are about 7 – 10 µm long, contain a single nucleus, are reproduced both by asexual (humans as well as other mammals) and sexual reproduction (cats) and form cysts for transmission.

**Disease:** Toxoplasmosis. Usually asymptomatic, it is an intracellular parasite of intestinal epithelial cells (first point of contact) as well as macrophages and muscle cells. But especially in the immunocompromised patients, it can cause fever, inflammation, swelling of the lymph nodes and disseminate to other organs including the lungs, liver, heart, brain and may cause death. Pregnant women with toxoplasmosis may pass it to their child, a form known as congenital toxoplasmosis.

**Diagnosis:** Diagnosis is made by direct demonstration of parasites and using serology-based assays.

**Transmission:** Through eating oocysts in cat faeces or by eating raw or uncooked meat of pigs, sheep, cattle, etc.

**Treatment:** Treatment is not necessary for healthy individuals but for immunocompromised or pregnant women it involves oral application of a combination of pyrimethamine and sulfadiazine.

**Occurrence:** Worldwide.

### *8.3.2 Organism: Cryptosporidium parvum*

**Biology:** The reproductive stage is about 7 µm, contain a single nucleus, reproduce both by asexual fission and sexual reproduction with male and female gametes and form oocysts for transmission.

**Disease:** Cryptosporidiosis (commonly occurs in immunocompromised patients). Often in AIDS patients, it infects intestinal epithelial cells causing watery diarrhoea lasting for several months. The frequency of bowel-movement ranges from 5 – 25 per day and could result in death, if not treated. In normal individuals, it may be asymptomatic or if it occurs, lasts a few days and is self-limiting.

**Diagnosis:** Diagnosis is made by direct demonstration of parasites in stool samples or serology-based assays.

**Transmission:** Oral uptake of *Cryptosporidium* oocysts from faeces with contaminated food or drinking water.

**Treatment:** The treatment involves oral application of nitazoxanide with limited effects.

**Occurrence:** Worldwide.

### *8.3.3 Organism: Cyclospora cayetanensis*

**Biology:** Similar to *Cryptosporidium* spp., are reproduced both by asexual fission and sexual reproduction with male and female gametes and form oocysts for transmission.

**Disease:** Usually asymptomatic but can produce watery diarrhoea with symptoms such as weight loss, abdominal pain, nausea and vomiting, fever and fatigue. Untreated infections typically last for 10 – 12 weeks and may follow a relapsing course. Infections, especially in disease-endemic settings can be asymptomatic.

**Diagnosis:** Diagnosis is made by direct demonstration of parasites in stool samples or serology-or PCR-based assays.

**Transmission:** Oral uptake of *Cyclospora* oocysts from faeces with contaminated food or drinking water.

**Treatment:** The treatment involves oral application of nitazoxanide with limited effects.

**Occurrence:** Worldwide.

### *8.3.4 Organisms: Isospora belli*

**Biology:** Oocyst containing sporocyst (18 – 30 μm) are taken up orally followed by sporozoite release and invasion of intestinal epithelial cells. Sporozoites divide asexually and produce merozoites, which invade epithelial cells and produce male and female gametocytes. Fertilization results in the development of oocysts that are excreted in the faeces.

**Disease:** Diarrhoea (mostly limited to immunocompromised patients). Usually in AIDS patients, it infects the intestinal epithelial cells causing heavy diarrhoea lasting for up to 24 months.

**Diagnosis:** Diagnosis is made by direct demonstration of parasites in stool samples.

**Transmission:** The faecal-oral route through food/water contaminated with *Isospora* oocysts.

**Treatment:** Treatment includes oral application of co-trimoxazole for couple of weeks.

**Occurrence:** Worldwide.

### *8.3.5 Organism: Plasmodium* spp.

**Biology:** Parasites differentiate into different forms in the vertebrate host and depending on the stage, the size varies from 2.5 – 15 μm, reproduce both by asexual and sexual reproduction.

**Disease:** Malaria. Infects the liver endothelial cells and the red blood cells. Symptoms include abdominal pain, headache and typical intermittent fever-chills and anemia associated with destruction of the red blood cells (periods vary depending on species). *Plasmodium falciparum* may produce continuous fever and results in death.

**Diagnosis:** Diagnosis is made by direct demonstration of parasites in stool samples and serology-based assays.

**Transmission:** The disease transmission is via insect bites, i.e., *Anopheles* mosquito (vector), as parasites are present in the salivary glands.

**Treatment:** Use of chloroquine, primaquine and qinghaosu in early disease is effective.

**Occurrence:** Warm countries (mostly tropical and subtropical countries).

## 8.4 Ciliates

### *8.4.1 Organism: Balantidium coli*

**Biology:** The trophozoite stage is about 50 – 60 µm long, contains a macro- and micronuclei, are reproduced by asexual binary fission and by conjugation and form cysts under harsh conditions as well as for transmission.

**Disease:** Balantidiasis. Symptoms include 80% of infections are asymptomatic and 20% infects intestinal epithelial cells producing ulcers. Advanced cases show symptoms similar to amoebic dysentery, i.e., vomiting, diarrhoea, nausea and could lead to death.

**Diagnosis:** Diagnosis is made by direct demonstration of parasites in stool samples.

**Transmission:** Faecal-oral route via food/water contaminated with *Balantidium* cysts.

**Treatment:** The treatment involves oral application of metronidazole, diiodohydroxyquin and tetracycline.

**Occurrence:** Worldwide.

## 8.5 Microsporidia

Microsporidia is a general term used to describe obligate intracellular protozoan parasites (> 1200 species). They produce resistant spores that infect both humans and animals. Spores associated with human infections are 1 – 4 µm, and possess a unique organelle, i.e., polar tubule coiled inside the spore. The human pathogens of the microsporidia include *Brachiola algerae, B. connori, B. vesicularum, Encephalitozoon cuniculi, E. hellem, E. intestinalis, Enterocytozoon bieneusi, Microsporidium ceylonensis, M. africanum, Nosema ocularum, Pleistophora* sp., *Trachipleistophora hominis, T. anthropophthera, and Vittaforma corneae. Enterocytozoon bieneusi* is the most common human pathogen causing 40% of all human microsporidian infections. Microsporidia mostly produce disease in AIDS patients. Various organs can be affected including the kidneys and intestines, and infectious spores are found in the urine and faeces. Symptoms of disease include diarrhoea and dysfunction of infected organs. Transmission: Oral uptake of spores from human urine or faeces with contaminated food, drinking water, hand-mouth contact or by unprotected homosexual intercourse. Diagnosis is made by direct demonstration of parasites or using serology or PCR-based assays.

Geographic distribution is worldwide.

**Treatment:** No satisfactory drug is available; albendazole and fumagillin have promising effects.

## 9. *BALAMUTHIA MANDRILLARIS* AS A MODEL PROTOZOAN

*Balamuthia mandrillaris* is an emerging opportunistic protozoan pathogen, a member of free-living amoebae. *Balamuthia mandrillaris* is known to cause serious cutaneous infections and fatal encephalitis involving the CNS with a case fatality rate of more than 98%. Despite such poor prognosis, the pathogenesis and pathophysiology associated with *Balamuthia* amoebic encephalitis (BAE) remain incompletely understood. Current methods of treatment include a combinatorial approach, where a mixture of drugs is administered, and even then the outcome remains extremely poor. There is an urgent need for improved antimicrobial chemotherapy and/or alternative strategies to develop therapeutic interventions. The purpose of this review is to discuss our current understanding of the biology and pathogenic mechanisms of *B. mandrillaris*.

### 9.1 Discovery of *B. mandrillaris*

*Balamuthia mandrillaris* was first isolated in 1986, from fragments of the brain tissue of a mandrill baboon (*Papio sphinx*) which died of a neurological disease at the San Diego Zoo Wild Animal Park in California, USA. The pathophysiological examination revealed that the animal died of a necrotizing haemorrhagic encephalitis similar to granulomatous amoebic encephalitis (GAE) caused by *Acanthamoeba*. Based on light and electron microscopic studies, animal pathogenicity tests, antigenic analyses, and nuclear (18S) and mitochondrial (16S) DNA sequences, a new genus, *Balamuthia*, was created for this amoeba, named after William Balamuth, a protozoologist who promoted the study of free-living amoebae, while the species designation indicates the source of the first isolate, i.e., a mandrill baboon. Later, in 1991, *B. mandrillaris* was associated with fatal human infections involving the CNS. Since then, more than 100 cases of BAE have been identified. At present there only is a single species in this novel genus, *Balamuthia mandrillaris*.

### 9.2 Classification of *Balamuthia mandrillaris*

When initially isolated from a mandrill baboon, *B. mandrillaris* was classified as a leptomyxid amoeba. Leptomyxidae (a group of soil amoebae) belong to the subclass, Lobosea which move using pseudopodia and exhibit characteristics such as limax, plasmodial, reticulated and polyaxial forms.

Limax refers to cytoplasmic flow, followed by a break where no movement is observed: such movement has been commonly attributed to leptomyxid amoebae. Plasmodial forms suggest that amoebae may possess more than one nucleus, while reticulate forms may resemble a network of fibres. The term 'polyaxial' refers to the ability of the amoebae to travel on more than one axis. Members of leptomyxid amoebae all have a thick cell walls in the cyst stage. With the observations of similar properties, *B. mandrillaris* was initially thought to be a member of Leptomyxidae. Lately, it has been acknowledged that morphology alone is insufficient for the classification of organisms and ribosomal ribonucleic acid (rRNA) has been suggested as a more accurate assessment in the classification of amoebae, as well as other organisms. However, Visvesvara's amoeba continued to be regarded as leptomyxid, until Stothard and co-workers determined the gene sequence of the nuclear small rRNA. These findings revealed that *B. mandrillaris* is the closest evolutionary relative to *Acanthamoeba*. Phylogenetic analyses supported a relationship between *B. mandrillaris* and *Acanthamoeba* through calculating the evolutionary distances. The authors suggested that *B. mandrillaris* should be removed from the Leptomyxidae family to the Acanthamoebidae family. These findings were further supported by findings that sequence dissimilarity between *B. mandrillaris* and *Acanthamoeba* strains is only between 17.9 – 21.1%. Based on these studies, it is currently accepted that *B. mandrillaris* is a close relative of *Acanthamoeba* and is placed in the family, Acanthamoebidae. Of note, the sequence variations in the genes of all *Balamuthia* isolates tested to date, range from 0 – 1.8%, thus all have been placed in a single species, *B. mandrillaris*.

## 9.3 Ecological Distribution

Generally speaking, free-living amoebae are widely distributed in the environment in a variety of habitats, including soil, water and even air. For example, *Acanthamoeba* and *Naegleria,* members of free-living amoebae have been isolated from a variety of environments. Although *Balamuthia* has been described as free living amoeba, there only are two reports of its isolation from the environment (i.e., soil). It is of interest that the isolation of environmental *B. mandrillaris* was from soil in potted plants. Such soil is often enriched organically with additives (chicken manure, earthworm castings, bat guano etc.), making it a rich environment for bacterial growth and, for organisms that feed on bacteria and one another. There are two reports of BAE in dogs, who swam in pond water, however, the isolation of *B. mandrillaris* from water samples remains undetermined. Despite limited success, the precise distribution, niche, or preferred food source of *B. mandrillaris* in the environment is not known. There may be several explanations for this, such as (i) *B. mandrillaris* are less abundant in the environment; (ii) are limited only to certain environmental niches; (iii) because they are difficult to isolate; (iv) that they are slow growing organisms (i.e., a life cycle of 20 – 50h), or a combination of the above. These are important questions and a complete understanding of them should help us design strategies to develop preventative measures for the susceptible hosts.

## 9.4 Isolation of *Balamuthia mandrillaris*

Unlike *Acanthamoeba*, which is widely distributed in the environment and can be easily isolated from a variety of settings, such as soil, water and even by air sampling, *B. mandrillaris* are fastidious to isolate and to culture. However, recently *B. mandrillaris* isolation from soil has been reported. Here, between 5 – 15 g of soil was suspended in 5 – 20 ml of sterile distilled water and a few drops were plated onto a 1.5% non-nutrient agar plate seeded with *Escherichia coli*. After many weeks, one plate exhibited amoebae of approximately 50 μm, which displayed irregular branching structures. Scrapings of the plates with amoebae resembling *B. mandrillaris* were transferred to tissue culture flasks containing monkey kidney cells as a feeder layer (although other mammalian cell cultures can be used). The identity of *B. mandrillaris* was confirmed using immunofluorescence staining assays and PCR methods. It is noteworthy that the overgrowth of fungi and other protists complicate the isolation of *B. mandrillaris*. For example in the aforementioned study, it took 10 – 20 transfer steps in order to separate the amoebae from the contaminating fungi. An additional problem is attributed to its slow growth, ranging from 20 to 50 h. The use of *Naegleria gruberi* to serve as an intermediate food source is shown to help to reduce an overwhelming growth of bacteria.

## 9.5 Axenic Cultivation

An enriched axenic medium for the axenic cultivation of *B. mandrillaris* has been established which contains Biostate peptone (2 g), yeast extract (2 g), Torula yeast RNA (0.5 g), liver digest (5%), MEM vitamin mixture, 100x (5 ml), MEM nonessential amino acids, 100x (5 ml), lipid mixture 1000x (0.5 ml), glucose (10%), hemin (2 mg/ml) and taurine (0.5%), Hank's balanced salts (10x) 34 ml, newborn calf serum 10% and the pH was adjusted to 7.2 with sterile 1*N* NaOH and final volume made up to 500 ml using distilled water. However, prolonged cultivation leads to loss of amoeba cytopathogenicity on tissue culture monolayers, while amoebae maintained on tissue culture cells, retain cytopathogenicity.

## 9.6 Storage of *Balamuthia mandrillaris*

*Balamuthia mandrillaris* trophozoites can be frozen for long-term storage. Briefly, log-phase amoebae (actively dividing) are resuspended at a density of $3 - 5 \times 10^6$ parasites per ml in freezing medium, containing 90% foetal bovine serum and 10% dimethyl sulfoxide. The cultures are transferred to –20°C for 4 h, and then at –80°C for 24 h, followed by storage in liquid nitrogen indefinitely. *Balamuthia mandrillaris* cultures can be revived by thawing at 37°C, followed by their immediate transfer to mammalian cells monolayer in a T-75 flask at 37°C. In our experience, the aforementioned freezing medium exhibits optimum revival viability.

## 9.7 Biology and Life cycle

The life cycle of *B. mandrillaris* consists of two stages; a vegetative trophozoite stage and a dormant cyst stage (Fig. 8). The trophozoites measure 12 – 60 μm in diameter and have a characteristic irregular branching structure. As a eukaryotic cell, *B. mandrillaris* possess a nucleus (approximately 5 μm in diameter) but may possess more than one nucleolus. Trophozoites reproduce asexually by binary fission, which is a form of mitosis where the nucleolus and the nuclear membrane divide to form daughter cells. At mitosis, the nuclear envelope remains intact with an intranuclear spindle until about metaphase, when the envelope breaks down and reforms at interphase. Centrioles are not seen at the poles, nor are they seen during the interphase. A microtubule-organizing centre (MTOC) is present. Other organelles include mitochondria, the organelles that are the sites of cellular respiration, the endoplasmic reticulum (ER) containing ribosomes, the site of protein synthesis, the Golgi apparatus, which is a membranous sac, acts to manufacture and produce complex glycoproteins. In old cultures, the average size of the amoebae decreases but upon inoculation on cell monolayers, the amoeba size returns to normal. Under adverse environmental conditions, such as lack of nutrients, extremes in pH or temperature, overcrowding of cells or excess of waste products, trophozoites transform into cysts: a process known as encystment. Encystment ensures the survival of the organism in adverse environmental conditions. Cysts tend to be smaller, i.e., 6 – 30 μm and are round in structure. Recent electron microscopic studies have shown that cysts in *B. mandrillaris* have three walls with a thin, irregular outer wall known as ectocyst, a fibrillar middle layer known as mesocyst and a thick inner wall called endocyst. So far, data on the longevity of cysts is lacking. Under favourable conditions, i.e., availability of nutrients, neutral pH, moderate temperature (30 – 37°C), cysts differentiate into trophozoite forms, a process known as excystment.

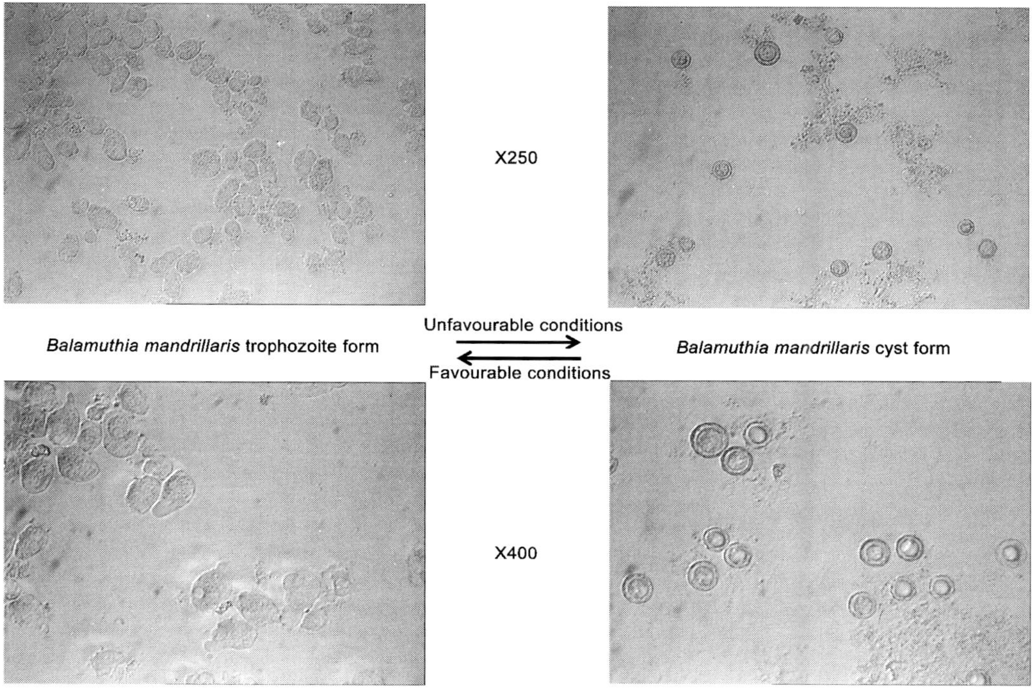


**Fig. 8** **The life cycle of *Balamuthia mandrillaris*.** The infective form of *B. mandrillaris*, also known as trophozoites, as observed under a phase-contrast microscope exhibiting distinct morphological characters as shown by arrows. Under harsh conditions, trophozoites differentiate into cysts. Although cysts are tripartite (when observed under an electron microscope), however under an optical microscope only two layers are observed.

## 9.8 Feeding (Prokaryotes, Single Cell Eukaryotic Organisms and Mammalian Cells)

Due to our incomplete understanding of the ecological distribution of *B. mandrillaris*, the preferential food source for these organisms remains unclear. Being a member of the free-living amoebae, it was thought that *B. mandrillaris* would feed on widely distributed prokaryotes. However, neither Gram-negative nor Gram-positive bacteria, supported *B. mandrillaris* growth, even though *B. mandrillaris* incubated with bacteria remained in the active trophozoites stage for more than 10 days, while amoebae incubated alone differentiated into cysts within 24 h. It was further shown that *B. mandrillaris* took up bacteria as demonstrated using fluorescein isothiocyanate (FITC)-labelled bacteria. Overall, these studies have demonstrated that *B. mandrillaris* are unable to consume bacteria for their growth but somehow utilize bacteria to remain in the trophozoites forms. Regardless, it is accepted that *B. mandrillaris* is a free-living amoeba which is distributed in the natural environment. In support, it is shown that normal populations possess antibodies against these potential pathogens. At least, one explanation of *B. mandrillaris* occurrence in the environment is thought to be due to their ability to feed on other protozoa (most likely amoebae), that also are widely distributed in the environment. Recent studies determined the use of *Acanthamoeba* as a potential food source for *B. mandrillaris*. It was observed that although *B. mandrillaris* exhibited growth on *Acanthamoeba* trophozoites, they showed a limited ability to target *Acanthamoeba* cysts. This is an interesting finding and may aid in determining the food selectivity in *B. mandrillaris*. For example, the cyst wall of *Acanthamoeba* is composed largely of polysaccharides (one third is cellulose), which is absent in *Acanthamoeba* trophozoites. The differential analysis of *Acanthamoeba* trophozoites and cysts may provide clues to identify the basis of discriminatory feeding behaviour of *B. mandrillaris*. In contrast, *B. mandrillaris* flourish on mammalian cells cultures. All tested cell cultures including, the human brain microvascular endothelial cells (HBMEC), human lung fibroblasts, monkey kidney (E6), African green monkey fibroblast-like kidney cells (Cos-7), rat glioma, mouse macrophage, and murine mastocytoma, supported the growth of *B. mandrillaris*. Overall, the food selectivity in *B. mandrillaris* may be dependent on the grazing properties of the amoebae and/or the susceptibility and biochemical properties of the prey and require further investigations.

## 9.9 *Balamuthia* Amoebic Encephalitis (BAE)

*Balamuthia* amoebic encephalitis is a chronic disease lasting from three months to two years, and almost always proves fatal. Unlike *Acanthamoeba* encephalitis, BAE has been found both in immunocompromised individuals and those with normal immunity, which indicates the virulent nature of this pathogen. Infection due to *B. mandrillaris* may be referred to as encephalitis or meningoencephalitis, depending on the degree of meningeal involvement. Of interest, meningoencephalitis involves the meninges surrounding the CNS, while encephalitis affects the brain itself, giving rise to typical symptoms such as seizures, cranial nerve palsies, and ataxia. The rarity of BAE suggests the presence of predisposing factor(s). Of note, BAE has been reported in patients suffering from diabetes, cancer, HIV-infected patients, or drugs and alcohol abusers. To date, more than 100 worldwide cases have been reported, but the exact number of cases of BAE may never be known and may be attributed to a lack of awareness, poor diagnosis and poor public health systems, especially in the less developed countries. Combinations of drugs are being used for the treatment with very limited success and this is of growing concern. A summary of the features of BAE is shown in Table 2.

**Table 2** Characteristics of *Balamuthia* amoebic encephalitis (BAE).

| | |
|---|---|
| **Susceptible hosts** | Immunocompromised hosts include HIV/AIDS patients, or individuals undergoing organ transplantation or steroid treatment, as well as drugs and alcohol abusers. Immunocompetent hosts usually include young children and older individuals. |
| **Symptoms** | Headache, fever, nausea, mental state abnormalities, irritability, hemiparesis, cranial nerve palsies, hallucinations, photophobia, sleep and speech disturbance, and seizures. |
| **Risk factors** | Breaks in the skin, working with soil without protective clothing. |
| **Incubation period** | Weeks to months. |
| **Cerebrospinal fluid** | High protein levels, low glucose levels, amoebae rarely found in the CSF. |
| **Gross pathology** | Multiple necrotic lesions in the brain. |
| **Histopathology** | Cysts and trophozoites found in the perivascular spaces, inflammation with or without granulomas. |
| **Prognosis** | Extremely poor, with more than 98% fatality rate. |

## 9.10 Portals of Entry

Several routes of entry into the CNS have been suggested. These may include amoebae penetration of the olfactory neuroepithelium via the nasal route. This was demonstrated in a mice model, where post-injection, amoebae were traced from the nasal epithelium to the submucosal surface, through the cribriform plate of the ethmoid bone along the olfactory nerve filaments, and then into the brain parenchyma. Furthermore, the gastrointestinal tract has also been shown to produce infection *in vivo* using mice. However, haematogenous dissemination from a primary lung (amoeba entry via respiratory tract), or skin (amoeba entry through breaks in the skin, followed by exposure to contaminated soil, e.g., working in gardens), is thought to be a more common route (Fig. 9). In support, it was observed that the lack of involvement of the olfactory lobes in *B. mandrillaris* infection, suggesting that the route of amoebae entry into the CNS was hematogenous, but the primary focus was not clear. Moreover, the parasites usually are found within localized areas of the brain and typically found around the blood vessels. Infections are often characterized by skin lesions, and in a few cases amoebae have been found in the kidneys, adreanal glands, pancreas, thyroid and lungs, and the fact that multiple tissues become affected suggest that hematogenous spread is an important step in BAE. Cutaneous infection may develop at the site of injury. Following the initial injury, it may take up to several months before developing into BAE. Amoebae entry into the CNS most likely occurs at the sites of the blood-brain barrier. However, recent studies showed the isolation of *B. mandrillaris* from the CSF of a BAE patient. Although it was widely accepted that amoeba may exist in the CSF, there had been a lack of evidence. This finding of *B. mandrillaris* in the CSF indicated that amoebae entry into the CNS may have occurred through the highly vascular choroids plexus.

## 9.11 Epidemiology

There have been over 100 reported cases of BAE. The majority of cases have been reported from Latin America i.e., 34 documented cases. Of these, BAE has been recorded in Peru (24 cases), Argentina (four cases), Brazil (one case), Mexico (four cases) and Venezuela (one case). The total number of reported cases in the USA is approximately 30. The majority of cases in the USA have been highlighted in the South-West, with California, Texas and Arizona, recording the most number of cases. In addition, BAE has been reported in Asia: Japan (two), Thailand (one), Australia (eight), and Europe: Czech Republic (one), and the United Kingdom (one). There is a predominance of cases in the very young (under 15 years of age) and the elderly (over 60 years of age), which may be attributed to somewhat weaker immune systems. However, *B. mandrillaris* has been found in both immunocompetent and immunocompromised individuals. An estimated of 25 cases of BAE have been reported in previously healthy, immunocompetent individuals. Furthermore, about half the cases in the USA have been reported in Hispanic Americans, even though they comprise about 13% of the population. But whether individuals of Hispanic origin are more exposed to *B. mandrillaris*, (exposure to soil due to socio-economic status and/or the fact that the majority of workforce in agriculture is of Hispanic origin in America) or whether they have a genetic predisposition

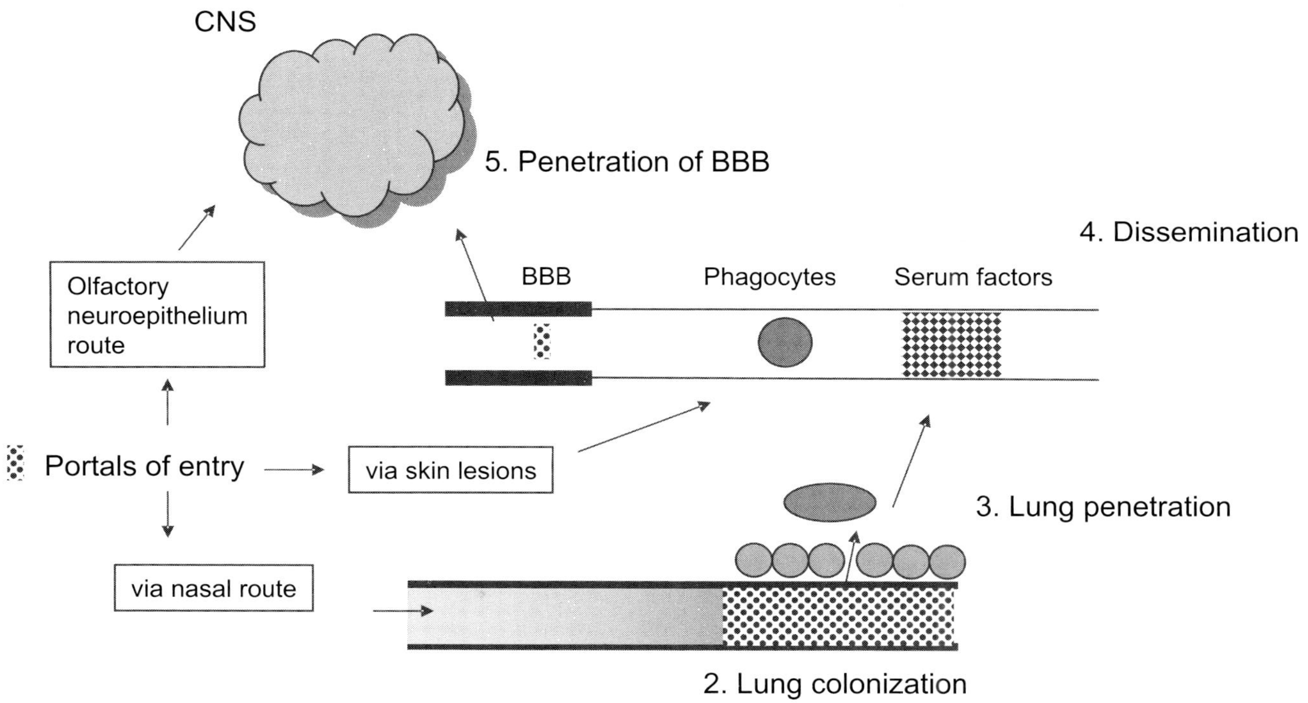


**Fig. 9** **The model of *Balamuthia* amoebic encephalitis.** The amoebae are thought to enter the lungs via the nasal route. Next, the amoebae traverse the lungs into the bloodstream, followed by haematogenous spread. Finally, *Acanthamoeba* cross the blood-brain barrier and enter into the central nervous system (CNS) to produce disease. It is noteworthy that *B. mandrillaris* may bypass the lower respiratory tract and directly enter into the bloodstream via skin lesions. The olfactory neuroepithelium may provide an alternative route of entry into the CNS.

to succumb to this disease remains undetermined. *Balamuthia* amoebic encephalitis is not only restricted to humans but have also been noted in other mammals including mandrill baboons, monkeys, gibbons, gorillas, sheep, dogs and horses.

## 9.12 Clinical Manifestation

After contracting *B. mandrillaris*, patients initially suffer from headache,a stiff neck, nausea and low-grade fever. As the disease progresses, patients become drowsy, there is a marked changes in behaviour, and speech may become incomprehensible. If the route of entry is breaks in the skin, painless nodules may be observed, which develop into lesions containing trophozoites whilst the patient is alive. Such lesions indicate a site of entry and are frequently observed in BAE patients. Symptoms may last from several weeks to months. Some patients exhibit hemi-paresis and weakness on part of the face or body, papilledema from increased intracranial pressure, hemiparesis, cranial nerve palsies (mainly of the third and sixth cranial nerves), as well as limitations in the patient's movement. The condition of the patient may further deteriorate with lack of response to stimuli together with pulmonary oedema or pneumonia, focal seizures, photophobia, and finally resulting in death. A summary of signs, features and symptoms of *B. mandrillaris* infections is indicated in Table 2. Gross pathological features including lesions may also develop in the brain accompanied by swelling i.e., (oedema) with the soft tissue and necrotic features such as haemorrhagic foci. Amoebae may also invade the parenchyma, though neurons are generally not affected. Microscopic examination reveals amoebic trophozoites and cysts found in perivascular spaces of brain sections and other infected tissues in patients. The infected brain shows characteristic granuloma formations with multinucleated giant cells, trophozoites and cysts. The granuloma formation is due to the inflammatory response of the host immune system.

## 9.13 Clinical Diagnosis

As *Acanthamoeba* encephalitis is mostly limited to immunocompromised patients, the susceptible hosts can be monitored for early signs of infection. However, the present data indicates that BAE is more difficult to detect, as it is rare and affects immunocompromised as well as immunocompetent individuals. The unpredictable nature of this disease means that BAE is even less likely to be diagnozed in time for medical intervention and, like *Acanthamoeba* encephalitis and primary amoebic meningoencephalitis due to *Naegleria fowleri*, it is essential for BAE to be diagnozed early if it is to be treated successfully. Clinical diagnosis is often made, once general symptoms have been exhibited by the patient. Due to the common features with other types of meningo-encephalitis, many cases of BAE infections are only diagnozed at the post-mortem stage. These infections are also frequently misdiagnozed and confused with brain tumours, abscesses, toxoplasmosis or cysticercosis. Lumbar puncture, which is the extraction of CSF from between the third and fourth vertebrae, is frequently used to determine the involvement of the CNS in infection. In the case of BAE, normal or low glucose levels, and slightly too high elevated protein levels are normally found (from ~30 mg/dl to > 1000 mg/dl), as well as the presence of white blood cells in the CSF (> 1000 cells $mm^{-3}$), which indicates the involvement of the CNS. In addition, magnetic resonance imaging (MRI) or computerized tomography (CT) are useful and exhibit brain lesions. Lesions are found in the parietal lobe, anterior lobe, temporal lobe, and cerebellum. In one case, there were as many as 50 lesions present in the brain. Lesions have been described as calcified and forming a mass-like structure. Biopsy may reveal features such as a necrotic cortex and the ghostly outline of perivascular monocytes. Despite these observations, to date, the prognosis remains poor with only three recorded cases of survival.

Once a person is suspected to have contracted BAE, laboratory diagnosis is the best way to confirm infection due to *B. mandrillaris*. This can be achieved using immunofluorescence methods and PCR-based assays. Successful laboratory detection of BAE has been described using immunofluorescence techniques. In immunofluorescence techniques, fluorescence microscopy can be used to detect the presence of antibodies towards a suspected target antigen of the disease-causing agent (antigen). A positive response is indicated by the specific excitation and emission of fluorescence. Polymerase chain reaction is developed for the specific diagnosis of *B. mandrillaris* using rRNA sequences. The PCR assays using chelex to isolate the DNA from *B. mandrillaris* could detect parasites in low cell numbers. This is important as often clinical samples may contain very few amoebae. Alternatively, *B. mandrillaris* can be described directly by histological examination, which requires expert knowledge of morphological characteristics of *B. mandrillaris*. Culture isolation from biopsies and/or CSF provides confirmatory evidence.

## 9.14 Predisposing Factors

Although BAE has been reported in immunocompetent populations (individuals who were negative for syphilis, diabetes mellitus, malignancies, fungal, HIV-1, HIV-2 and mycobacterial infections as well as contained normal CD4- and CD8-postive T-lymphocytes counts and B-lymphocytes), however given the rarity of the disease it is hard to imagine that there are no predisposing factors in contracting BAE. But whether the predisposing factors are other primary infections, underlying genetic factors, mere exposure to the environment with widely distributed *B. mandrillaris*, or a combination of the above is not completely understood. For example, exposure to contaminated soil has been a major risk factor in contracting BAE. In one case, a Californian man working in his backyard developed an infection soon after sustaining a puncture wound that was probably contaminated by soil, and, in a second case, a woman from New York was reported to have worked in her garden with compost soil prior to developing an infection both of whom were immunocompetent. Based on these and others it appears that exposure to contaminated soil is a major risk factor. In addition, the victims of BAE have included immunocompromised patients, as well as alcoholics, drug users and persons with diabetes, undergoing excessive antimicrobial chemotherapy, steroid treatment, radiotherapy, or organ transplantation appear to be more susceptible.

## 9.15 Prevention and Control

A significant number of BAE infections have been associated with individuals who had contact with soil. It has been suggested that immunocompromised people are most at risk from contracting BAE infection, but healthy individuals with no underlying disease may also be at risk of contracting this infection. People who are agricultural workers or have regular contact with soil may be more susceptible to BAE infection. Once in the CNS, *B. mandrillaris* is difficult to treat and control, unless identified at a very early stage of infection. Clearly, prevention of this infection from happening in the first instance is the most effective solution. Certain simple precautionary measures may be taken to prevent *B. mandrillaris* from entering the host. Individuals should avoid working in soil without protective clothing especially if there are skin lesions present. For example, in one case a child was found to have contracted BAE infection after playing with soil from a potted plant, which may be an additional risk factor.

Although *B. mandrillaris* has not as yet been isolated from water, people who are immunocompromised should be warned of the risks of swimming in freshwater streams, rivers and lakes, not just for *B. mandrillaris*, but also for other free-living pathogens.

## 9.16 Antimicrobial Therapy for BAE

Presently there is no specific or even recommended treatment against BAE and the case fatality rate is more than 98%. A limited rate of success depends on the early diagnosis followed by aggressive treatment. Due to the lack of knowledge regarding BAE treatment, clinicians have used a mixture of antimicrobials with the hope for a successful outcome. However, such combinational therapy may have adverse side effects on the patient. To date, successful treatment has only been achieved in three patients. In the first case combinational therapy of pentamidine (300 mg intravenously once a day); sulfadiazine (1.5 g, orally, four times a day), fluconazole 400 mg once daily and clarithromycin 500 mg, three times a day, was administered. Other successful outcomes have been observed through treatment using flucytosine (2 g four times a day, orally), pentamidine isethionate (4 mg/ kg/ day intravenously), sulfadiazine (1.5 g, four times a day), fluconazole (400 mg/ day) for two weeks, followed by treatment with clarithromycin (500 mg/ day). However, the use of pentamidine was discontinued in both cases cited, because of the side effects. In addition, pentamidine had poor penetration into the CNS when tested in HIV/AIDS patients. The diamidines with improved ability to cross the blood-brain barrier and with low toxicity would be highly desirable in treating amoebic and other CNS infections. A major concern in these two cases was for recurrence of the infection as a result of reactivation of dormant cysts in the brain lesions. However, there was no evidence of reactivation of the disease, but the patients remain on fluconazole plus sulfadiazine or clarithromycin. Other drugs are currently being tested as a potential means of combatting BAE infection. Preliminary studies indicate miltefosine (hexadecylphosphocholine) as a potential drug for treating BAE. Miltefosine is an alkylphosphocholine drug previously used for treatment against protozoan diseases, such as leishmaniasis and is suggested to cross the blood-brain barrier. It is an enzyme inhibitor and is well absorbed when taken orally through the human body. Miltefosine has been tested against clinical isolates of *B. mandrillaris* using different concentrations of the drug, showing effective amoebacidal and amoebastatic properties, *in vitro*, at 30 – 40 µM concentration. Despite this, the majority of BAE cases have been fatal. Even with antemortem diagnosis of BAE, initiation of effective antimicrobial therapy may come too late to help the patient. The use of flucytosine, fluconazole, pentamidine, and clarithromycin at regular or high dosages was ineffective in four Argentinean pediatric BAE cases. Future studies should identify novel drugs and/or determine the potential of known compounds with increased blood-brain-barrier permeability, in the treatment of BAE.

## 9.17 Pathogenesis of BAE

Pathogenesis of a disease refers to the ability of a parasite to bring about the disease. There is a lack of information regarding the pathogenesis of *B. mandrillaris*. In addition, the ability of *B. mandrillaris* to produce encephalitis in immunocompetent individuals as well as those who are immunocompromised is a matter for concern, and indicates the virulent nature of this pathogen. As indicated earlier, portals of entry possibly include breaks in the skin and/or respiratory tract followed by amoebae invasion of the intravascular space. Haematogenous spread is thought to be a key step in BAE, but it is not clear as to how circulating amoebae cross the blood-brain barrier to gain entry into the CNS to produce the disease. In blood, *B. mandrillaris* is subjected to the immune system of the host, which involves leukocytes, macrophages, neutrophils and the complement pathway. In addition, antibodies are present in the normal populations against *B. mandrillaris*, with levels of 1:32 to 1:256 titre in human sera. However, it is shown that despite the presence of highly efficient immune systems, *B. mandrillaris* can produce infection in immunocompetent individuals. The crossing of the blood-brain barrier is indeed thought to be a critical step which is required by many CNS pathogens. First observed in 1885, the blood-brain barrier is a highly selective barrier which restricts the entry of toxins/pathogens into the CNS due to the presence of tight junctions. The blood-brain barrier is thus critical in the pathogenesis of the CNS disease, when it fails to prevent an invading pathogen through its selective mechanism. The blood-brain barrier separates the blood from the interstitial and CSF of the CNS. The blood-brain barrier can be further divided at two levels (Fig. 10), (a) a barrier between the blood and brain extracellular fluid located at the cerebral capillary endothelium, and (b) a barrier between the blood and CSF located at the choroid plexus epithelium and the arachnoid membrane as discussed below.

### 9.17.1 Blood-brain barrier at the choroid plexus (a barrier between the blood and the cerebrospinal fluid, CSF)

The choroid plexus is highly vascularized epithelium found within each of the four cerebral ventricles and is responsible for the production of CSF (Fig. 10). Here, endothelial cells are fenestrated to allow penetration of the blood contents but are surrounded by epithelial cells of the choroid plexus. Like capillary endothelium, the polarized choroid plexus epithelium also form tight junctions (Fig. 10), however these exhibit lower electrical resistance compared with the brain capillary endothelium.

### 9.17.2 Blood-brain barrier at the brain parenchyma (i.e., a barrier between the blood and the brain tissue)

Tight junctions of the cerebral capillary endothelium prevent entry of blood contents into the CNS, making it a highly selective barrier compared with the peripheral endothelium. For comparison, endothelial cells of the blood-brain barrier exhibit extremely high transendothelial electrical resistance (> 1,000 ohm per $cm^2$), while endothelial cells from human placenta exhibit < 50 ohm per $cm^2$). In addition to endothelial cells, blood-brain barrier is composed of basement membrane (containing laminin, collagen IV,

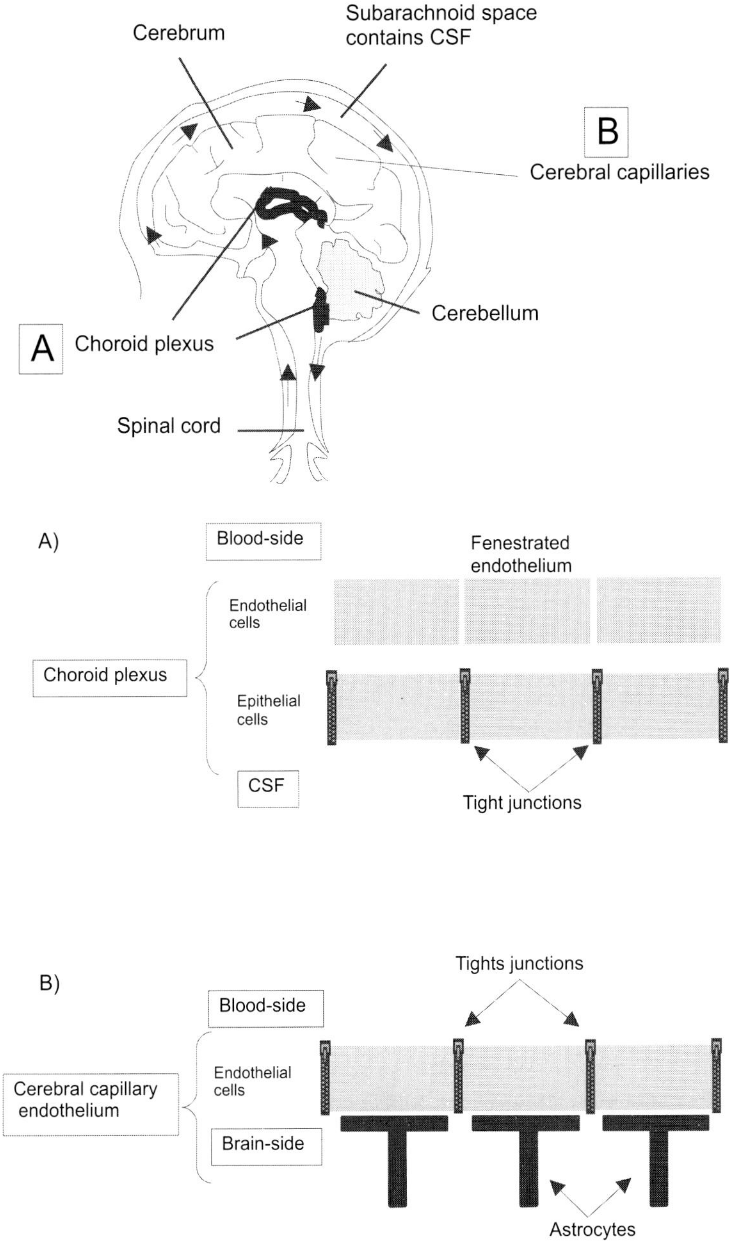


**Fig. 10** Based on locations, the blood-brain barrier can be further divided at two levels, (A) a barrier between blood and cerebrospinal fluid (CSF) located at the choroid plexus epithelium and (B) a barrier between blood and brain extracellular fluid located at the cerebral capillary endothelium. Arrows indicate the flow of cerebrospinal fluid within the CNS.

proteoglycans (especially heparan sulphate), fibronectins, nidogen, and entactin), astrocyte (a type of glial cells) that extend foot-like projections around the outside of the capillaries, and pericytes embedded within the basement membrane (Fig. 11). In particular, astrocytes are important in the development and maintenance of the blood-brain barrier. Pericytes are non-neural cells that are thought to regulate the capillary blood flow. Pericytes and astrocytes are surrounded by the basal lamina, over which lies the extracellular matrix. The presence of tight junctions in the blood-brain barrier ensures that even small molecules such as dyes, and antibiotics are prevented from entry into the CNS by limiting the paracellular route. In contrast, non-brain endothelium is more permissive. However, the small lipophilic molecules substances such as $O_2$ and $CO_2$ can diffuse freely across the plasma membrane along their concentration gradient. The nutrients such as glucose, amino acids, etc., are transported across the blood-brain barrier via transporters, while larger molecules such as insulin, leptin, iron transferrin are taken up via receptor-mediated endocytosis to support the neuronal function. Junction complex in the blood-brain barrier comprise of tight junctions and adherens junctions (Fig. 11).

**9.17.2i Tight junctions** These consists of three integral membrane proteins (i.e., claudin, occludin, junction adhesion molecule), and cytoplasmic proteins [zonula-1 (ZO-1), ZO-2, ZO-3] (Fig. 11). Claudins are 22 kDa phosphoproteins with four transmembrane domains that bind to claudins on the adjacent endothelial cells to form tight junctions. The carboxy terminal of claudins binds to cytoplasmic zonula proteins. Occludin is a 65 kDa phosphprotein with four transmembrane domains and its cytoplasmic domain is directly associated with zonula proteins. Junctional adhesion molecules (JAM) are 40 kDa with a single transmembrane domain. Zonula proteins (ZO-1) is 220 kDa, ZO-2 is 160 kDa, and ZO-3 is 130 kDa are accessory proteins that provide structural support. They are shown to bind to all three of the above. The cytoplasmic proteins further link membrane proteins to actin cytoskeleton.

**9.17.2ii Adherens junctions** Are composed of membrane protein, cadherin that bind to catenin. The cytoplasmic domains of cadherins bind to catenin which are linked to actin cytoskeleton and form adhesive contacts between the cells. Adherens junctions include cadherin, actinin and vinculin (analogue of catenin). Both tight junctions and adherens junctions are present in epithelial cells of the choroid plexus. Claudins-1, -2, -11, occludin and ZO-1 are present in choroid plexus epithelial cells (blood-CSF barrier), whereas claudins-1, -5, -11, occludin and ZO-1 are present in the blood-brain barrier.

An *in vitro* model of the blood-brain barrier using HBMEC has been previously developed. These cells are positive for factor VIII related antigen

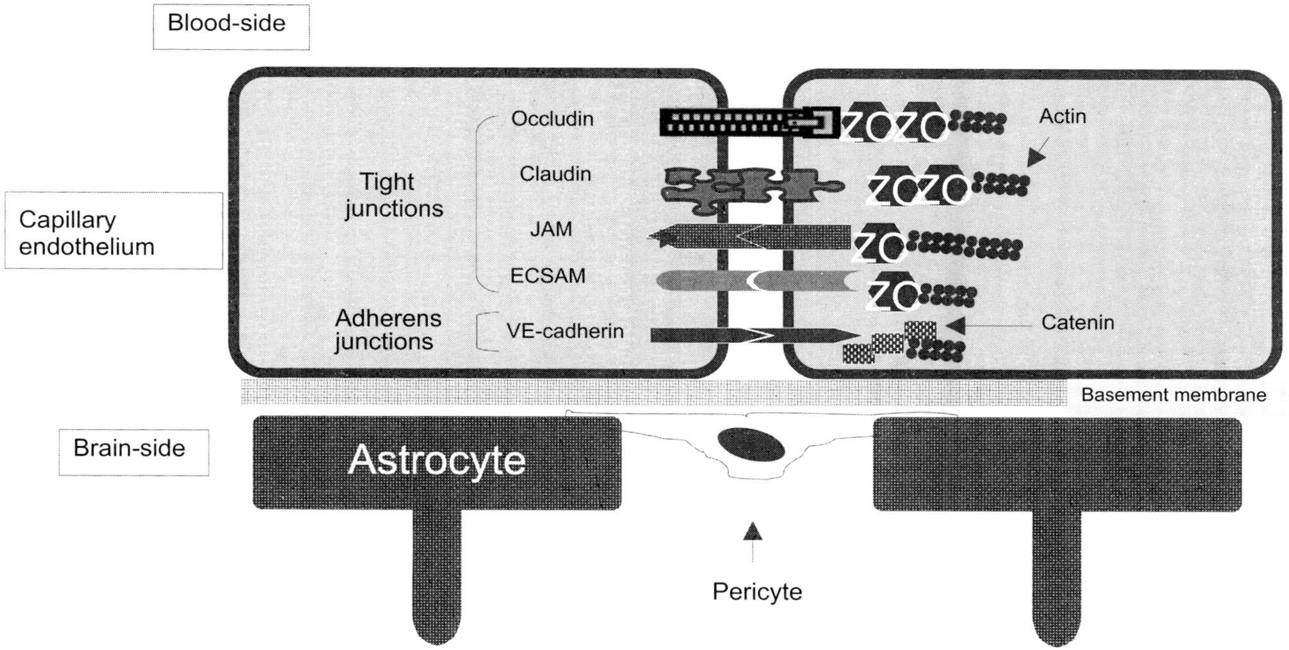


**Fig. 11 Schematic illustration of the blood-brain barrier at the cerebral capillary endothelium exhibiting tight junctions.** The endothelial cells are surrounded by basement membrane, which is ensheathed by astrocytes and pericytes. The tight junctions are composed of integral proteins including occludin, claudin, junctional adhesion molecule (JAM), and endothelial cell-selective adhesion molecule (ECSAM), that interact with their counterparts on the adjacent endothelial cells. Both tight junctions and adherens junctions are composed of multiple proteins. The cytoplasmic tails of these proteins interact with the actin cytoskeleton via a number of accessory proteins including members of zonula-occludens (ZO).

and carbonic anhydrase IV, which indicates their endothelial origin, as well as positive for gamma-glutamyl transpeptidase indicating their brain origin. Thus, it provides a physiologically relevant model to study *B. mandrillaris* traversal of the blood-brain barrier. To this end, we utilized HBMEC as a model of the blood-brain barrier and studied their interactions with *B. mandrillaris* (Fig. 12).

In studies related to many other pathogens causing meningitis or encephalitis, the pathogen must traverse the blood-brain barrier to produce the disease. There may be three mechanisms of an amoeba crossing the blood-brain barrier. The first involves receptor-mediated transport, a contact-dependent mechanism, whereby the amoebae adhere to endothelial cells via an adhesin. In related protozoa, for example, *Acanthamoeba,* mannose binding protein is the adhesin involved in interactions with the blood-brain barrier. The second mechanism may involve a paracellular route, whereby the amoebae traverse the blood-brain barrier by crossing in between endothelial cells at the tight junctions. The third involves direct crossing by producing damage to the endothelium. Given the size of *B. mandrillaris* (15 – 65 μm), it is most likely involves penetration of the barrier. A number of events such as adhesion of amoebae to HBMEC, cell injury and inflammatory response may combine to disrupt the blood-brain barrier.

## 9.18 Inflammatory Response to *B. mandrillaris*

The main purpose of an inflammatory response is to recruit leukocytes to the site of infection and to allow the host to counteract invading agents. The inflammatory response involves cytokines such as, tumour necrosis factor-alpha (TNF-α), interleukin-1 (IL-1), interleukin-6 (IL-6), interleukin-8 (IL-8) and granulocyte/ macrophage colony stimulating factor (GM-CSF). Cytokines are non-structural proteins that originate from lymphocytes and monocytes, and possess a number of functions that include mediation of signalling communications, tissue repair, and inflammation. Cytokines are produced by different cells and may exert various effects under differing circumstances. Cytokines may work synergistically with other cytokines or act as agonists. In particular, among the pro-inflammatory cytokines to be released is interleukin-6 (IL-6) which is a pleiotropic cytokine that plays a key role in initiating early inflammatory responses. Interleukin-6 has several biological activities in the body, such as the modulation of T and B cell functions and acute phase reactions. Interleukin-6 is known to play a definite role in the pathogenesis of a number of infections. For example, interleukin-6 antibodies have been found to attenuate inflammation in a rat model for bacterial (pneumococci) meningitis. In endothelial cell lines, there is evidence linking IL-6 to increased permeability of the blood-brain barrier by reducing transendothelial resistance and shear stress as well as

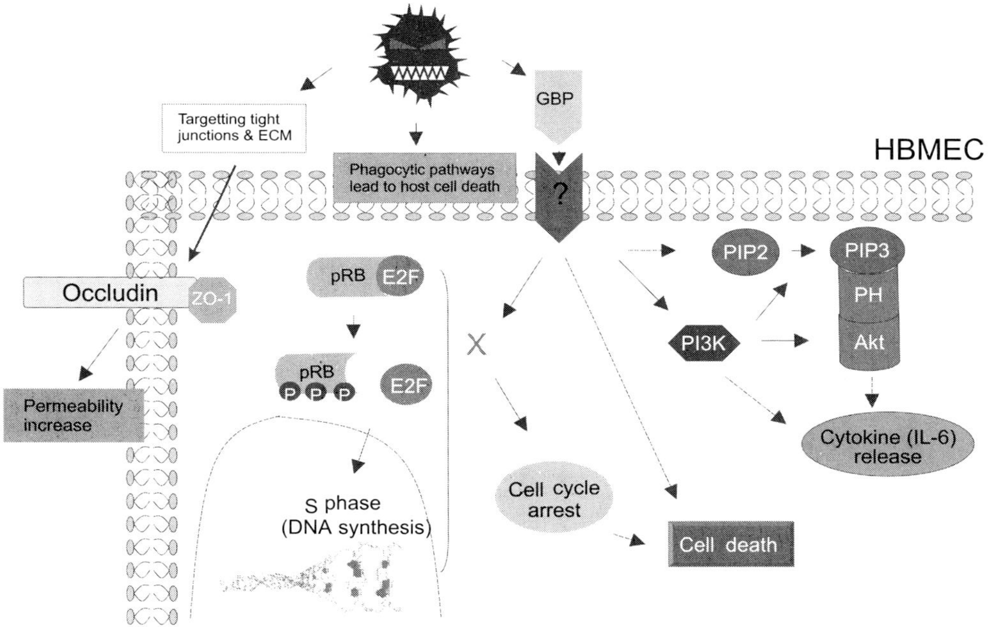


**Fig. 12 Host intracellular signalling in response to *B. mandrillaris*.** Note that *B. mandrillaris* induces cell cycle arrest in the host cells by (i) altering expression of genes as well as by (ii) modulating protein retinoblastoma (pRb) phosphorylations. In addition, *B. mandrillaris* have also been shown to induce inflammatory response by inducing the release of interleukin-6 from HBMEC, via phosphatidylinositol 3-kinase (PI3K) and produce host cell death via the phagocytic pathways. The amoebae disrupt tight junctions by targetting zonula-1 and occludin proteins. GBP is galactose-binding protein; E2F is a transcription factor that controls cell proliferation through regulating the expression of essential genes required for cell cycle progression; PIP2 is phosphatidylinositol-4,5-bisphosphate; PIP3 is phosphatidylinositol-3,4,5-trisphosphate; Akt (protein kinase B)-PH domain, a serine/threonine kinase is a critical enzyme in signal transduction pathways involved in cell proliferation, apoptosis, angiogenesis, and diabetes.

by modulating the expression of a number of adhesion molecules on endothelial cells in order to encourage leukocyte infiltration of the blood-brain barrier. Interleukin-6 exerts its effect by binding to the cell surface receptor complexes, which belong to the type I cytokine receptor family. These consist of two types of receptors which include non-signalling α-receptors such as, IL-6-Rα where R indicates the receptor and secondly, signal-transducing receptors such as glycoprotein 130 (gp130). Glycoprotein 130 receptor subunit was identified and cloned as the signal transducer for IL-6. When activated, the gp130 activates janus kinases tyrosine kinase (JAK). This in turn activates multiple signal-transduction pathways, including the STATs (signal transducer and activator of transcription), Ras MAPK (mitogen activated protein kinase) and phosphatidylinositol 3-kinase (PI3K) pathways. These ultimately can lead to cell damage, cell death or the induction of protective pathways mediated by the activation of nuclear transcription factor kappa B (NF-κB). We determined that *B. mandrillaris* induced IL-6 production by HBMEC, which may play a role in *B. mandrillaris* traversal of the blood brain-barrier. Other cytokines such as TNF-α were tested, but did not elicit any remarkable findings. The role of PI3K in *B. mandrillaris*-mediated IL-6 release in host cells was investigated by using LY294002, an inhibitor of PI3K, which reduced IL-6 release, confirming the role of PI3K. Furthermore, Western-blotting assays confirmed the role of Akt, a known downstream effector of PI3K, indicating that phosphorylation of Akt required the activation of PI3K. The use of dominant-negative cells containing a plasmid containing a mutant regulatory subunit of PI3K confirmed our findings. Here, HBMEC which expressed dominant-negative PI3K displayed differences in *B. mandrillaris*-mediated IL-6 release, compared to the vector alone (pcDNA3) which showed normal levels of IL-6 release. It was concluded that *B. mandrillaris* mediated IL-6 release in HBMEC was dependent on PI3K activation. Due to possible transcriptional activities, it is possible that the NF-κB may play a role in this signalling pathway. The Akt has also been linked to the regulation of transcriptional activity of NF-κB by phosphorylating the p65 NF-κB subunit. Future studies are needed to determine the role of PI3K in NF-κB activation and its involvement in IL-6 release in response to *B. mandrillaris*. The model of HBMEC may also be used in future studies to investigate other inflammatory processes associated with *B. mandrillaris*-mediated response in the host. Of interest would be nitric oxide release, which has been associated with IL-6 release and the blood-brain barrier permeability. A full cytokine profile, using a gene expression array could determine specific cell signalling pathways involved in inflammatory response, apoptosis and/or necrosis.

## 9.19 *Balamuthia mandrillaris* Adhesion to the Blood-brain Barrier

Recent studies have shown that *B. mandrillaris*-mediated host cell death is a contact-dependent mechanism. It was further shown that separation of the amoebae from target cells by a semipermeable membrane prevented host cell destruction, as did lysates of the amoebae and inhibitors of actin polymerization. Using HBMEC, which constitute the blood-brain barrier we have recently shown that *B. mandrillaris* produces HBMEC cytotoxicity, which may lead to the blood-brain barrier perturbations. However the underlying molecular mechanisms associated with amoebae traversal of blood-brain barrier leading to pathological features remain unclear. Although the successful traversal of *B. mandrillaris* across the blood-brain barrier may require multiple events, adhesion is a primary step in amoebae transmigration of the HBMEC. Our recent studies showed that *B. mandrillaris* binds to HBMEC in a galactose-inhibitable manner and identified a galactose-binding protein (GBP) expressed on the surface of *B. mandrillaris* of approximate molecular weight of 100 kDa. The presence of GBP in *B. mandrillaris* has recently been suggested by demonstration that *B. mandrillaris* binds to laminin and these interactions can be inhibited using exogenous galactose. Our results support these findings and have identified the expression of a GBP on the surface membranes of *B. mandrillaris*. However, given the complexity of host-parasite interactions, it is tempting to speculate that the aforementioned interactions only provide initial attachment, which is most likely followed by closer associations with a more intimate contact of *B. mandrillaris* to HBMEC involving GBP, as well as other adhesin(s). Such binding is probably necessary to withstand blood flow as well as for subsequent crossing of the blood-brain barrier. Of interest, GBP inhibited *B. mandrillaris*-mediated HBMEC cytotoxicity. Further studies are needed to validate the concept of other determinants, in addition to GBP and their roles in *B. mandrillaris*-HBMEC interactions.

## 9.20 Phagocytosis

The fast and efficient killing of the host cells by *B. mandrillaris*, suggests the involvement of an active killing mechanism rather than metabolic poisoning or nutrient depletion by the amoebae. Binding leads to secondary events such as interference with host intracellular signalling pathways, toxin secretions, and the ability to phagocytose host cells, ultimately leading to cell death. Our studies suggested that interactions of *B. mandrillaris* with host cells stimulate specific host cell signalling pathways, resulting in the host cell death. This is confirmed at the protein level by studying the phosphorylation of the retinoblastoma protein (pRb), a master regulator of the cell cycle (Fig. 13).

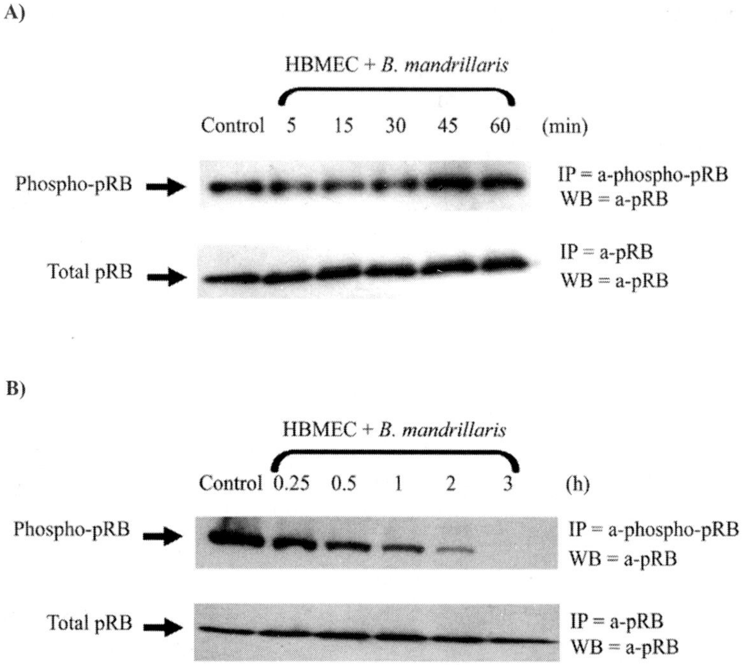


**Fig. 13** ***Balamuthia mandrillaris*-induced protein retinoblastoma (pRB) dephosphorylation in human brain microvascular endothelial cells (HBMEC).** The HBMEC were grown in 60 mm dishes and incubated with *B. mandrillaris* (approx. $2 \times 10^6$ amoebae) for up to 60 min (A) or up to 3 h (B). Proteins were immunoprecipitated (IP) with anti-phospho-pRb antibodies and immunoblotted with anti-pRb antibody. In controls, proteins were immunoprecipitated and immunoblotted with anti-pRb antibody. Note that *B. mandrillaris*-induced pRb dephosphorylation in HBMEC in a time-dependent manner.

Both scanning and transmission electron microscopy revealed extensive morphological changes of the amoebae during feeding, as the target cell is not only enveloped but also penetrated. In axenic cultures, *B. mandrillaris* trophozoites appear in a large variety of shapes, ranging from spherical, almost smooth, to highly polymorphic with intense surface activity. Both spherical and polymorphic forms can bind to multiple mammalian cells at a time by, as yet, unknown means and then quickly begin to engulf and penetrate their targets, further supporting the concept, that target cell lysis and phagocytosis are intimately connected. A prominent amoebastome, as it is characteristic of similarly cytopathic *Acanthamoeba* spp., is also shown to present in *B. mandrillaris*.

## 9.21 Ecto-ATPases

Ecto-ATPases are present on the external surface of *B. mandrillaris*. The external localization of the ATP-hydrolyzing site is supported by their sensitivity to suramin, which is a non-competitive inhibitor of ecto-ATPases and an antagonist of P2 purinoreceptors, which mediate the physiological functions of extracellular ATP. We observed that live *B. mandrillaris* hydrolyze extracellular ATP using *in vitro* assay. The ability of *B. mandrillaris* to hydrolyze ATP may have a role in the biology and pathogenesis of *B. mandrillaris*. Moreover using non-denaturing PAGE, we identified an ecto-ATPase of molecular mass of > 595 kDa which is heat-stable but labile to detergent, SDS. In addition, it was insensitive to ouabain, levamisole, sodium azide and sodium orthovanadate, further confirming that ATP hydrolysis is due to ecto-ATPase. Previous studies have shown that ADP release from *A. castellanii* plays an important role in its contact-independent cytotoxicity, as demonstrated by increased levels of intracellular calcium, which subsequently leads to apoptosis in Wish cells. This is further supported by the inhibition of *Acanthamoeba*-mediated host cell cytotoxicity using ecto-ATPase inhibitor, suramin (P2 receptor antagonist), clearly demonstrating that ecto-ATPases play an important role in the pathogenesis of *Acanthamoeba*. Being closely related to *Acanthamoeba*, it is reasonable to predict that *B. mandrillaris* may employ similar mechanisms to produce host cell damage. In support, we determined that suramin an ecto-ATPase inhibitor, blocked > 40% *B. mandrillaris* binding to the host cells. The role of ecto-ATPase in host-parasite interactions is further supported by findings that suramin blocks *B. mandrillaris*-mediated HBMEC cytotoxicity.

Considering the fact that *B. mandrillaris* is responsible for brain infection, whose ulceration of the tissues allows the parasite to penetrate into the host CNS, one could speculate that the presence of ecto-ATPase activity might reflect some form of evasion of the parasite from the host defence mechanisms in the circulation. For example, a highly active ecto-ATP-diphosphohydrolase was localized on the external surface of the tegument of *Schistosoma mansoni*, and it was suggested that this enzyme might be involved in an escape mechanism allowing the parasite to split ATP, released by activated cytotoxic T lymphocytes. A similar role may be attributed to *B. mandrillaris* ecto-ATPase; however it remains to be determined. In conclusion, the detection of an ecto-ATPase in *B. mandrillaris* is an important step to understand the possible role of this enzyme in the pathogenesis of BAE, which may help us to identify potential targets to intervene these serious infections.

To produce damage to the host cells and/or tissue migration, the majority of pathogens rely upon their ability to produce hydrolytic enzymes. These enzymes may be constitutive which are required for routine cellular functions or inducible, that are produced under specific conditions, for example upon contact with the target cells. These enzymes can have devastating effects on host cells by causing membrane dysfunction or physical disruptions. Cell membranes are made of proteins and lipids, and other free-living amoeba, *Acanthamoeba* is known to produce hydrolytic enzymes: proteases, which hydrolyze peptide bonds; and phospholipases, which hydrolyze phospholipids. Recent studies have shown the presence of phospholipase A, lysophospholipase A, and lipase activities in *B. mandrillaris*, which may have a role in the host cell membrane degradation. However, their specific roles in target cells lysis remain to be established.

In addition, recent studies have shown that *B. mandrillaris* exhibited phospholipase $A_2$ and phospholipase D activities in a spectrophotometric-based assay. The functional role of phospholipases was determined using HBMEC *in vitro*. It was observed that $PLA_2$-specific inhibitor i.e., cytidine 5′-diphosphocholine partially inhibited *B. mandrillaris* binding to HBMEC. Similarly, PLD inhibitor i.e., compound 48/80 inhibited *B. mandrillaris* binding to HBMEC. Moreover, both inhibitors partially blocked *B. mandrillaris*-mediated HBMEC cytotoxicity. Overall these results clearly demonstrate that phospholipases may play important roles in the pathogenesis of *B. mandrillaris*.

## 9.22 Proteases

During the entry into CNS, from the primary skin lesion or nasal epithelium amoebae encounter molecules of the host extracellular matrix (ECM), including components of the basal lamina. In healthy brains, ECM comprises a major per cent of the normal brain volume, which forms the basal lamina around the blood vessels. The ECM is constantly remodelled and provides the critical structural and functional support, as well as maintaining homeostasis in the neuronal tissue. However, in neurological disease states, ECM may undergo substantial modifications resulting in neuroinflammatory responses. The excessive ECM degradation affects neurovascular structural/functional properties that are highly destructive to the CNS functions. The ECM is composed of both collagen types and non-collagenous glycoproteins and the proteoglycans. Thus, the ability of the amoebae to degrade ECM may aid in their invasion of and growth in the brain tissue. Recent studies have shown that *B. mandrillaris* bind to the extracellular matrix components, laminin, collagen and fibronectin. Collagen is the primary component of the ECM and is difficult to degrade due to its helical structure. Our recent studies have shown that *B.*

*mandrillaris* exhibit protease properties and are able to cleave type I and III collagen at neutral pH, suggesting that *B. mandrillaris* proteases may play a role in ECM destruction. This was further confirmed using metalloprotease inhibitor, i.e., 1, 10-phenanthroline which completely abolished the protease activities. Moreover, *B. mandrillaris* metalloproteases exhibited elastinolytic activities. This was demonstrated by the degradation of elastin as substrate in SDS-PAGE. Previous studies have shown that elastase destruct ECM, which increases the blood-brain barrier permeability resulting in the brain injury. For example, injection of elastase into the CSF opened the blood-brain barrier in newborn piglets. In addition to the aforementioned, the urokinase plasminogen activator system plays an important role in various neuronal diseases involving the CNS inflammation and/or pathology. For example, in bacterial meningitis, the uPA (urokinase-type plasminogen activator) or tPA (tissue-type plasminogen activator) are known to convert plasminogen (abundant in brain), into plasmin, which destructs ECM directly by degrading fibrin or by activating matrix metalloproteases. We have shown that *B. mandrillaris* degraded proenzyme, plasminogen suggesting that pathogenesis of BAE may involve uPA/tPA system. Again, it has been shown that CSF injection of plasmin results in increased capillary permeability in rats. Further studies are in progress to determine the activation of plasminogen in response to *B. mandrillaris* metalloproteases and their targets in the neuronal tissue. In tandem, these mechanisms allow *B. mandrillaris* to target zonula-occludens-1 (ZO-1) and occludin (Fig. 14). As indicted above both ZO-1 and occludin are involved in the formation of tight junctions, suggesting that *B. mandrillaris* disrupt tight junctions to induce HBMEC permeability and/or HBMEC monolayer disruptions, leading to blood-brain barrier perturbations, and finally amoebae entry into the CNS to produce the disease.

## 9.23 Indirect Virulence Factors

The ability of *B. mandrillaris* to produce human diseases is a multifactorial process and amongst other factors, is dependent on their ability to survive outside its mammalian host for various times and under diverse environmental conditions (high osmolarity, varying temperatures, food deprivation, and resistance to chemotherapeutic drugs). The ability of *B. mandrillaris* to overcome such conditions can be considered as contributory factors towards disease and are indicated as indirect virulence factors.

## 9.24 Immune Response to *B. mandrillaris*

*Balamuthia mandrillaris* antibodies of the IgG and IgM classes were detected in healthy populations with titers ranging from 1:64 to 1:256. Cord blood

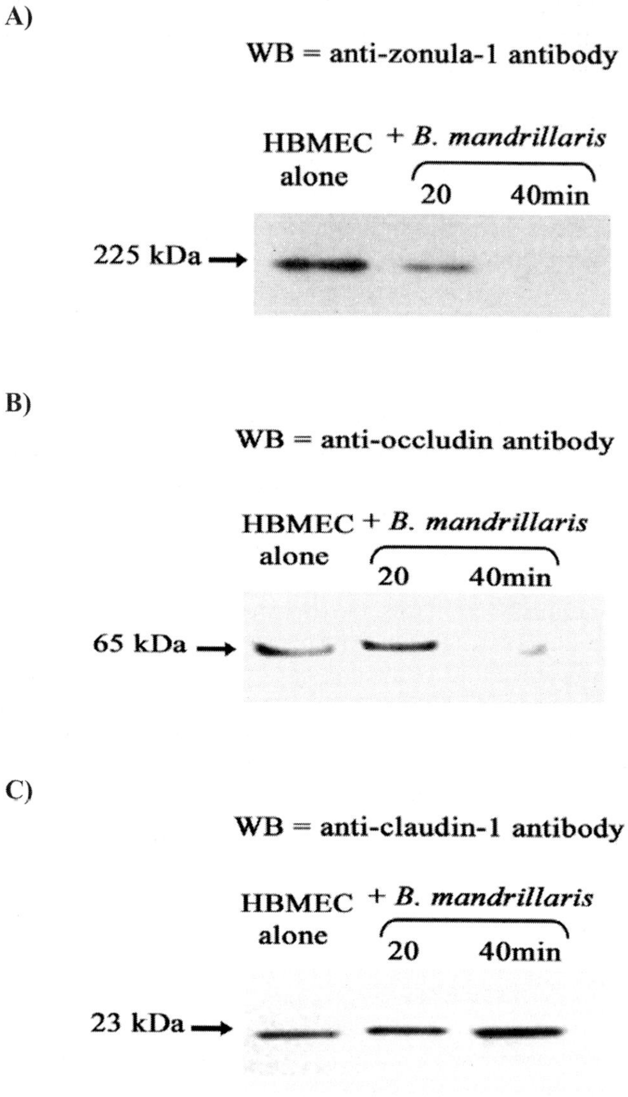


**Fig. 14** ***Balamuthia mandrillaris*-induced perturbation of the tight junctions barrier.** *Balamuthia mandrillaris* were incubated with the confluent human brain microvascular endothelial cell (HBMEC) monolayers for various intervals of time and cells were lysed by RIPA lysis buffer. Equal amounts of cell lysates were used for Western blotting assays using anti-ZO-1 (A), anti-occludin and (B) anti-claudin antibodies. (C) Note that *B. mandrillaris*-induced loss of ZO-1 and occludin within 40 min, but had no effect on claudin-1. Results are representative of three independent experiments.

also contained antibodies but at lower titers, perhaps the result of cross-placental transfer from the maternal circulatory system. However, the antibody levels were very low in neonates, which suggest that these substantially increase with age, probably as a result of environmental exposure to the amoeba in the soil.

Our recent study showed that serum exhibited protective effects on *B. mandrillaris* binding to and subsequent cytotoxicity of HBMEC. The normal human serum exhibited an initial limited amoebacidal effects, approx. 40% trophozoites were killed. However, a sub-population of amoebae remained viable but cultures were stationary over longer incubations. The fact that serum exhibited approx. 50% inhibition of amoebae binding to HBMEC (similar to amoebacidal effects) suggests that effects of serum on the properties of *B. mandrillaris* are at least partly secondary to the amoebacidal/amoebastatic effects. This is consistent with previous findings, which showed that virulent strains of *Acanthamoeba* resists serum-mediated killing. One interesting finding was that serum possess antibodies which reacted with several *B. mandrillaris* antigens in Western blotting assays. The antigens of *B. mandrillaris* reacted strongly with normal human serum. *Balamuthia mandrillaris* isolated from a baboon tissue (ATCC 50209) and from the human brain tissue shared several common antigens and confirmed that both isolates are antigenically close and belonged to the same species. Overall these studies suggest that normal human serum is partially adept of inhibiting *B. mandrillaris* properties associated with its pathogenesis but whether a healthy immune response is sufficient to control and/or eradicate this life-threatening pathogen is unclear. To this end, studies are being conducted to determine the detrimental effects of serum on *B. mandrillaris* in the presence of neutrophils/macrophages. These studies should clarify the mechanisms associated with *B. mandrillaris* pathogenesis, which may help design preventative measures and/or develop therapeutic interventions. However, the protective role of antibodies against BAE is somewhat unclear. For example, several BAE patients were reported to possess high titre of anti-*B. mandrillaris* antibodies without a protective response and resulted in death. This may be due to delayed humoral response, overwhelming BAE infection, or the ability of amoebae to evade the humoral immune response.

## 9.25 *Balamuthia mandrillaris* as a Host

There is a single report demonstrating that *B. mandrillaris* can serve as a host for *Legionella pneumophila*. The bacteria remained and multiplied within large vacuoles inside the amoeba. The continued incubation resulted in rounding and detachment of host amoeba resulting in their disintegration.

## 9.26 Conclusions

*Balamuthia* amoebic encephalitis is a chronic disease that almost always proves fatal. Unlike *Acanthamoeba* granulomatous encephalitis, BAE has been noted both in immunocompromised individuals. The identification of GBP as a major adhesin for *B. mandrillaris* is important, in that, it may identify novel targets for the rationale development of therapeutic interventions. This is not a novel concept. For example, in *Acanthamoeba*, it is shown that the oral immunization with rMBP protected animals against *Acanthamoeba* keratitis, *in vivo*. Similar strategies may be developed against BAE and the identification of GBP should lay a foundation for additional studies. Overall, future research should continue to identify the precise mechanisms associated with the pathogenesis of BAE, as well as host susceptibility, *B. mandrillaris* colonization of the skin lesions, and/or nasopharynx, leading to amoebae entry into the intravascular space, parasites survival within the bloodstream, invasion of the CNS and the brain tissue damage leading to encephalitis, which may help identify potential targets for the rationale development of therapeutic interventions and/or design preventative strategies against BAE.

CHAPTER

6

# Fungi

## 1. INTRODUCTION

Fungi are eukaryotic organisms that contain more than 300,000 species and some (few hundred) have been recognized as human pathogens, causing mild infections to serious life-threatening diseases. The majority of fungal pathogens cause opportunistic infections, i.e., normally limited to individuals with a weak immune system. The increasing numbers of immunocompromised patients especially due to HIV have made a large part of the human population susceptible to the fungal infections.

Similar to protozoa, the study of mycology began with the discovery of microscope. They were first observed by Robert Hooke in 1667, followed by Leeuwenhoek in 1689 observing their structure and budding. Later Antonio Micheli (1729) described hundreds of fungi as independent organisms. Thus far, the majority of fungi identified were plant pathogens and were identified from the decaying matter. For the next 100 years, the major focus in the field of mycology remained in their role as plant pathogens. However, in 1839, fungus (*Trichophyton*) was identified as the causative agent of human disease causing cutaneous infections.

## 2. TAXONOMY

As indicated previously, all livings beings are divided into two domains (Prokaryota and Eukaryotes). Within Prokaryota, there are two kingdoms, Archaebacteria and Eubacteria (Bacteria). Eukaryota is composed of five kingdoms including Protozoa, Chromista, Plantae, Animalia and Eumycota. Here, fungi are classified into two domains, Chromista and Eumycota, which are further divided into 5 phyla as follows,

## 2.1 Chromista (also called Fungi Imperfecti or Deuteromycota)

1. Phylum Hyphochytriomycota
   Fungi exhibit asexual reproduction and no known sexual stage. Fungi have a haploid vegetative stage and possess cellulose-containing cell walls. These include fungi that are mostly unknown to humans.
2. Phylum Oomycota (also called water moulds)
   Fungi exhibit asexual reproduction and no known sexual stage. Fungi have a diploid vegetative stage and possess cellulose-containing cell walls.

## 2.2 Eumycota

1. Phylum Chytridiomycota
   Fungi undergo sexual (produce oospores) and asexual reproduction. These fungi are most predominant in causing diseases in agriculture and in lower cold-blooded animals.
2. Phylum Zygomycota
   Fungi undergo sexual (produce zygospores) and asexual reproduction. Zygomycota include food spoilage agents.
3. Phylum Dikaryomycota (contain 90% of all fungi).

3a. Subphylum Ascomycetes – Fungi undergo sexual (produce ascospores) and asexual reproduction.

3b. Subphylum Basidiomycetes – Fungi undergo sexual (produce basidiospores) and asexual reproduction.

# 3. FUNGI – CELLULAR PROPERTIES

Fungi are eukaryotes and contain following structures:

1. Capsule
   Capsules are present only in some fungi, e.g., *Cryptococcus neoformans* (encapsulated yeast). Capsules are made of polysaccharides. They may play a role in the pathogenesis by inhibiting phagocytosis, thus can be considered as virulence factors.
2. Cell wall
   Cell walls are made of polysaccharides (approx. 90%) and approx. 10% of proteins and glycoproteins that provide skeletal support. In eumycotan fungi, cell walls are largely composed of chitin (made of N-acetyl-D-glucosamine, GlcNAc monomers), while in chromistan fungi, the cell wall is made of cellulose (made of D-glucose monomers). Other polysaccharides found in fungi are chitosan (polymers of D-glucosamine), $\alpha$-glucan (polymers of D-glucose), $\beta$-glucan (polymers of D-glucose), mannan (polymers of D-mannose), galactans (poly-

mers of D-galactose) and small amounts of fucose, rhamnose, xylose and uronic acids.

Their functions reside in providing shape, rigidity, strength and protection from osmotic shock. They are antigenic in nature (i.e., induce antibody response).

2. Cell membrane

   The cell membrane is made of a lipid bilayer. In contrast to other eukaryotes, the cell membrane of fungi contains ergosterol or zymosterol instead of cholesterol. The functions of cytoplasm include regulating material/fluid diffusion in and out of the cell as well as active transportation and facilitating capsule and cell wall synthesis.
3. Cytoplasm containing nucleus with chromosomes, nuclear membrane, nucleolus, endoplasmic reticulum, mitochondria, Golgi apparatus, vacuoles, ribosomal RNA and a cytoskeleton with microtubules composed of tubulin, intermediate filaments and microfilaments composed of actin. Their genome size is considerably smaller than other eukaryotes. For example, the genome size of *Saccharomyces cerevisiae* is approx. 13.5 Mbp as compared to *E. coli* that is around 5 Mbp, *Drosophila* is 165 Mbp and humans is 2900 Mbp. Apart from chromosomal genes, they also possess mitochondrial DNA, plasmids and mobile genetic elements. Similar to other eukaryotes, their genes contain non-coding regions, introns but shorter (around 100 bp) than other higher eukaryotes (around 10 kbp).

## 4. FEEDING

Fungi obtain their food by secreting hydrolytic enzymes, digesting their food externally and absorbing it. They obtain their food using three mechanisms.

1. Saprophytes – in this group, fungi use dead organic materials such as plants, animals, other microbes, etc.
2. Symbiosis – fungi live with their living host to the benefit of the host.
3. Parasitism – fungi utilize living tissues of plants and animals causing their damage.

Their need for carbon sources is met by oxidation of sugars, alcohols, proteins, lipids and polysaccharides to generate energy for their energy needs.

### 4.1 Carbon Nutrition

The hyphae absorbs sugars through their rigid cell wall in the soluble forms by diffusion. Once through the cell wall, fungi actively uptake sugars across their plasma membrane. However, sugars may not be available as soluble forms especially in the case of saprophytes and may be present as cellulose, chitin, etc. Here, fungi use degradative enzymes (i.e., cellulases, chitinases) to break down complex sugars into six carbon sugars such as glucose, followed by their uptake. Once taken up, carbohydrates (e.g., glucose) undergo glycolysis in the cytoplasm. During glycolysis, glucose is phosphorylated at the expense of two ATP molecules and results in fructose-1,6-diphosphate. It is then converted to two 3C compounds which are then converted to two molecules of pyruvic acid. Under aerobic conditions, pyruvic acid molecules enter a mitochondrion and are converted to two 2C molecules of acetyl-coenzyme A (CoA) with the release of two molecules of carbon dioxide. Acteyl-CoA is then combined with oxaloacetate (4C) to form citric acid (6C). Citric acid molecules undergo a series of reaction of the tricarboxylic acid cycle (TCA) to metabolize to oxaloacetate. Citric acid to oxaloacetate cycle runs twice for the complete oxidation of glucose molecule to generate energy with the net result of 38 molecules of ATP.

$$C_6H_{12}O_6 + 6O_2 \rightarrow 6CO_2 + 6H_2O + 38ATP$$

The majority of fungi are aerobic but some (e.g., yeast) can metabolize sugars under anaerobic conditions. Under oxygen-limiting conditions, puruvic acid molecules are converted into ethanol (with the aid of alcohol dehydrogenase) or lactic acid (with the aid of lactic dehydrogenase), which is released into the surrounding medium. The majority of yeast and mycelial fungi produce ethanol (e.g., *Saccharomyces* spp.) and several chytridiomycota produce lactic acid (e.g., Allomyces). This form of energy production is called fermentation, however this is an inefficient process and two molecules of ATP are generated for each glucose molecule and thus a large amount of sugar is needed for fungal growth. Fungi can also use non-sugar substrates as carbon source.

## 4.2 Nitrogen Nutrition

Fungi do not fix nitrogen. As with complex sugars, the complex nitrogen sources are taken up following their break down from the environment using degradative enzymes, i.e., extracellular proteases. Their nitrogen needs are met from nitrate, nitrite, ammonium, organic nitrogen, which is used for amino acid synthesis for proteins, purines and pyrimidines for nucleic acid synthesis, glucosamine for chitin and other needs. However, fungi are able to synthesize lysine using biosynthetic pathway.

Other growth factors such as phosphorous, potassium, magnesium, calcium, sulphur, copper, manganese, sodium, zinc and other requirements are normally available to fungi in the environment. However, fungi also produce phosphatases to enable them to access complex phosphate stores.

Other nutritional requirements include water, hence fungi usually grow in moist or damp environments. Fungi can grow at a variety of temperatures and pH.

## 5. GROWTH IN FUNGI

Fungal growth involves two stages, a vegetative stage and a reproductive stage.

1. Vegetative stage – during this stage, cells are normally haploid and divide mitotically. Most fungi exist as multicellular moulds with hyphae but some fungi exist as unicellular yeast cells. Some fungi can change their morphology and exists as both, and are called dimorphic. For example, *Candida* is found in the yeast form at 37°C but changes to the mould form at 25°C. The dimorphic property of fungi is regulated by temperature, $CO_2$ concentration, pH, levels of cysteine, etc.
2. Reproductive stage – Fungi may undergo sexual or asexual reproduction. Asexual reproduction involves the generation of spores, while sexual reproduction requires specific structures that are also used for taxonomic differentiation.

### 5.1 Reproductive Stage

In general, Eumycota fungi grow both by asexual and sexual reproduction while chromista fungi have only been shown to undergo asexual reproduction.

#### *5.1.1 Sexual reproduction*

Sexual reproduction is achieved by the formation of spores (gametes) that are haploid and fuse together to produce diploid zygote. Sexual spores are oospores, zygospores, ascospores and basidiospores that have been useful in the taxonomic differentiation. Zygote formation is usually followed by asexual reproduction to produce spores and thus completing the cycle.

#### *5.1.2 Asexual reproduction*

In asexual reproduction, fungi produce spores. Both eumycota and chromista reproduce in the form of spores or conidia. However, they are haploid in all fungi except oomycota that are diploid. Different fungi produce different spores including arthrospores, blastospores, chlamydospores, macroconidia, microconidia, sporangiospores. They are usually thick-walled which are normally produced by budding and are formed. They are nonmotile and present as single-cells or present in multi-compartment sac. Both spores and conidia are resistant to drying and harsh conditions.

Upon favourable conditions, spore germinates in the form of a germ tube (hypha approx. 4 to 9 μm wide). Hyphae walls are largely composed of chitin in the eumycotan fungi and cellulose in the chromistan fungi that grow rapidly at its tip. The interior of the hypha may be continuous or perforated at regular intervals called septa that strengthen the hypha and may regulate nuclear migration. Hyphae grow at the tip and branch extensively creating a network called mycelium.

#### *5.1.3 Parasexual reproduction*

Alternatively, hyphae fuse together, followed by nuclear material fusion or genetic exchange.

## 6. FUNGAL TRANSMISSION

The spores act both as mechanisms to survive harsh conditions as well as a way to disperse to more favourable environments. The spores produced by sexual processes (i.e., oospores, zygospores, ascospores) do not germinate readily upon contact with favourable conditions. They must mature or require specific conditions (appropriate temperatures, nutrition, moisture, pH, etc). While spores that are produced following asexual reproduction (i.e., sporangiospores, coniodiospores) are solely produced to disperse to other favourable environments. These spores are produced in large masses and disperse by wind or by water currents. It is also helped with the fact that spores are usually small and light in weight and thus can travel far. Other mechanisms of dispersal include explosion of the spore-containing hypha, allowing the transmission for maximum distance or dispersal using insects or animals as vectors. In addition to transmitting to favourable environments, the presence of spores in the air may have serious implications for human health. The favourable conditions allow the germinations of spores.

## 7. STRATEGIES AGAINST PATHOGENIC FUNGI

### 7.1 Chemotherapy

As with protozoan pathogens, chemotherapeutic approaches remain the most common form of treatment of human fungal diseases. These provide the most direct and cheapest way of controlling these infections. Many of the currently available antifungal drugs are produced synthetically or obtained from other organisms/materials. Fungal cell membranes possess ergosterol instead of cholesterol as in humans. The inhibitors of ergosterol biosynthesis such as azole compounds, allylamine and morpholine are widely used anti-fungal drugs. In addition, 5-fluorocytosine (DNA and

RNA synthesis inhibitor) or griseofulvin (mitosis) are important anti-fungal compounds. However, a complete understanding of fungal metabolism and their biosynthesis should help identify novel targets for the rational development of drugs. As with other microbial agents, it is paramount to combine other control measures in addition to chemotherapeutic approaches to help ease the sufferings of infected individuals.

## 7.2 Control Measures

### *7.2.1 Physical measures*

As indicated above, fungal pathogens form spores that are frequently present in the air or on surfaces and abundant in the soil. However, these fungi can not enter the keratinized skin to access the susceptible tissue. Even the minor skin lesions or wounds provide an excellent portal of entry for fungal (especially yeast) pathogens. Thus, any cuts should be cleaned with ethanol and protected to avoid contact with pathogenic fungi.

### *7.2.2 Sanitary measures*

The maintenance of good sanitary measures such as draining swamps, building sewage systems, and most importantly providing clean water supplies will reduce fungal reservoir in the environment and is crucial for controlling fungal infections. Avoidance of exposure to the environments favourable to fungal growth. Control of primary infections reduces the risks and incidence of fungal infections. For immunocompromised patients (due to HIV or organ transplantation), the use of antifungal compounds as prophylactic approaches may be crucial in the control of fungal infections.

# 8. HUMAN FUNGAL INFECTIONS

Of the vast number of known fungi, only a few evoke a pathogenic human response (called mycoses), notable exceptions being the species *Cryptococcus neoformans, Candida albicans, Blastomyces dermatitidis, Coccidiodes* spp. and *Histoplasma capsulatum.*

## 8.1 Organism: *Sporothrix schenckii*

**Biology:** It is a dimorphic yeast-like fungus, and is around 3 – 5 μm in diameter. The yeasts are not encapsulated.

**Disease:** It causes Sporotrichosis, which initiate with the direct inoculation into the skin resulting in lesions which develop into painless nodules that may ulcerate and drain. Very rarely pulmonary or systemic infections can occur in immunocompromised patients.

**Diagnosis:** Requires culture from the clinical biopsy or serology-based assays.

**Treatment:** The treatment includes the use of Amphotericin B or itraconazole.

**Transmission:** They are commonly present in plants or plants materials and are transmitted to humans via contact and rarely inhalation.

**Occurrence:** Worldwide.

## 8.2 Organism: *Blastomyces dermatidis*

**Biology:** Dimorphic fungi (i.e., exists as mould at 25°C and as yeast at higher temperatures). The yeast is thick-walled and measures 8 – 15 μm in diameter (occasionally up to 30 μm). Mould form exhibit blastoconidium.

**Disease:** Once inhaled, conidia transform into yeast form and within a month may cause a pneumonia-like pulmonary disease or remain asymptomatic and cause infection of lungs, skin, bones, genitourinary tract or other organs.

**Diagnosis:** Culture from the clinical biopsy.

**Treatment:** The treatment includes the use of Amphotericin B. Other alternative drugs include itraconazole, ketoconazole or fluconazole.

**Transmission:** They are commonly present in moist soil with enriched organic materials and are transmitted via inhalation of the conidia.

**Occurrence:** Worldwide.

## 8.3 Organism: *Coccidioides immitis*

**Biology:** It is a dimorphic fungus, which form endospores 2 – 5 μm in diameter, that are released in the environment.

**Disease:** Once arthroconidia are inhaled, they form spherules containing endospores and produce coccidioidomycosis (lung infetion) which may invade into other tissues including the CNS (usually in immunocompromised patients) with fatal consequences.

**Diagnosis:** Diagnosis requires a combination of clinical symptoms and culturing of the pathogen or the use of DNA probes.

**Treatment:** If the disease is limited to the lungs then treatment is not necessary but in severe cases the use of ketoconazole, amphotericin B or fluconazole are effective.

**Transmission:** They are usually found in dust, alkaline soils, deserts and transmitted to humans via inhalation.

**Occurrence:** Worldwide

## 8.4 Organism: *Histoplasma capsulatum*

**Biology:** It is a dimorphic fungus. The yeast is around 3 μm with a characteristic polysaccharide capsule.

**Disease:** Causes infection of the lungs (usually asymptomatic) but in immunocompromised patients, it may invade other tissues with fatal consequences.

**Diagnosis:** Diagnosis requires isolation of fungus or detection of fungal antigen (capsule).

**Treatment:** If the disease is limited to the lungs then treatment is not necessary but in severe cases, this includes the use of itraconazole, ketoconazole or amphotericin.

**Transmission:** They are usually found in the soil and transmitted to humans via inhalation.

**Occurrence:** Worldwide.

## 8.5 Organism: *Candida albicans*

**Biology:** *Candida* are thin-walled, yeasts approx. 4 to 6 μm that reproduce by budding.

**Disease:** They can cause infections of the skin and moist wet mucosal surfaces (cutaneous candidiasis, e.g., mouth, vagina), nail infections (candidal onchomycosis) and in immunocompromised patients they can cause invasive candidiasis, which may involve the heart, kidneys, genitourinary tract, liver, eye, bones, the CNS or other organs.

**Diagnosis:** Diagnosis requires a combination of clinical symptoms and culturing of the pathogen.

**Treatment:** Skin and cutaneous candidiasis can be treated with application of antifungal agents such as nystatin or amphotericin, while itraconazole, fluconazole, imidazole are also effective. Antifungal agents, itraconazole, fluconazole, amphotericin B and 5-fluoocytosine are effective against invasive candidiasis.

**Transmission:** They are commonly found in the soil, inanimate objects, food, and hospital environments.

**Occurrence:** Worldwide.

## 8.6 Organism: *Cryptococcus neoformans*

**Biology:** It is ubiquitous, encapsulated yeast-like fungus and is around 4 to 6 μm with a characteristic polysaccharide capsule.

**Disease:** It can cause infections of the lungs but in immunocompromised patients, it may produce meningitis (cryptococcal meningitis), that is inflammation of the membranes covering the brain and the spinal cord with serious consequences.

**Diagnosis:** Diagnosis requires a combination of clinical symptoms and culturing of the pathogen or detection of fungal antigen (capsule).

**Treatment:** The treatment includes the use of antifungal, amphotericin B, fluconazole or 5-fluorocytosine.

**Transmission:** They are usually found in soil to humans via inhalation.

**Occurrence:** Worldwide.

## 8.7 Organism: *Pneumocystis carinii*

**Biology:** This has been controversial in its classification and has been considered as a protozoan (since it reproduces from a cyst to trophozoite). But recent ribosomal RNA analyses suggests that it is more closely related to fungus than to protozoa and hence included here as a fungus. Trophozoites are about 1 – 5 μm long, contain a single nucleus, are reproduced both by asexual and sexual reproduction and form cysts under harsh conditions as well as for transmission.

**Disease:** It causes pneumonia (limited to immunocompromised patients) by infecting the alveolar epithelial cells in lungs. Symptoms include fever, cough and breathing problems and pathogen may disseminate to spleen, lymph nodes, bone marrow and may result in death if not treated.

**Diagnosis:** Diagnosis is made by demonstration of parasites in the sputum or lung tissue.

**Transmission:** Through aerosol droplets, sputum or direct contact with infected individuals.

**Treatment:** Pentamidine, or combination of trimethoprim-sulfamethoxazole.

**Occurrence:** Worldwide.

## 8.8 Organism: *Aspergillus* spp.

**Biology:** They form long, branching, septate hyphae that are approximately 3.0 µm in diameter which form conidiophores structures containing conidia.

**Disease:** Normally cause secondary infections. Depending on the host immune status, they can cause infection of the sinuses and lungs called allergic bronchopulmonary aspergillosis and pulmonary aspergilloma, and invasive aspergillosis which may lead to the CNS aspergillosis, osteomyelitis, endophthalmitis, endocarditis, renal abscesses, cutaneous infections.

**Diagnosis:** Culture from the clinical biopsy or demonstration of fungal antigens using commercially available EIA kits is recommended.

**Treatment:** Allergic aspergillosis can be treated with corticosteroids, and intraconazole. For Invasive aspergillosis, the use of voriconazole, amphotericin B or itraconazole are effective.

**Transmission:** They are commonly present in the soil and trasmitted to humans via inhalation of the conidia.

**Occurrence:** Worldwide.

## 8.9 Fungal Agents of Cutaneous Mycoses (also called Dermaotophytoses or Tinea and Ringworm Diseases)

These include fungi that cause infections of the hair, nails and skin. They are caused by three genera, *Epidermophyton* spp., *Microsporum* spp. and *Trichophyton* spp. The infecetions are normally limited to keratinized tissues. *Epidermophyton* spp. normally attack the skin, *Microsporum* spp. attack hair and skin but not nails and *Trichophyton* spp. target hair, skin and nails. Fungi are acquired through contact with infected individuals.

# 9. *CRYPTOCOCCUS NEOFORMANS* AS A MODEL FUNGUS

*Cryptococcus* was first isolated in 1894 by Busse, a pathologist who observed round-to-oval structures in a sarcoma-like tumour of the tibia of a woman. Subsequently *C. neoformans* was isolated mainly from meningoencephalitis patients, however the disease remained very rare. However, since 1980s, there has been a dramatic increase in the number of cases with the increasing numbers of AIDS patients. AIDS is now known to be a predisposing factor for the disease. *Cryptoccocus neoformans* is now said to be the chief progenitor of cryptococcal meningoencephalitis infections affecting 7 – 10% of those with AIDS.

## 9.1 Serotypes and Varieties

*Cryptococcus neoformans* is a yeast-like fungus belonging to the family Cryptococcaceae and reproduces by budding. The cell is usually around 4 – 6 µm in diameter with eukaryotic features as well as 5S, 18S and 25S rRNA. They normally form smooth and yellowish colonies on solid culture medium. They can be differentiated from the non-pathogenic species of *Cryptococcus* by growth at 37C (only *C. neoformans*) grow at 37°C). *Cryptococcus neoformans* possess a capsule, however the structure, and morphological characteristics of the capsule may vary depending on growth conditions. Based on their capsular polysaccharide, there are four serotypes of *C. neoformans* including A, B, C and D. The majority of the clinical isolates of *C. neoformans* belong to serotype A and D (although serotype A are significant higher in environmental and clinical isolates as compared to serotype D). Serotype A and D can mate (sexual reproduction) to produce a sexual state, which is known as *Filobasidiella neoformans* var. *neoformans*. The sexual state of serotype B and C are classified as *Filobasidiella neoformans* var. *gattii*. Both *C. neoformans* var. *neoformans* and *C. neoformans* var. *gattii* differ in ecology, epidemiology and their biochemical needs. For example, serotype B and C isolates (i.e., *C. neoformans* var. *gattii*) assimilate malic acid, fumaric acid, succinic acid and glycine as a sole carbon source, while serotype A and D isolates (i.e., *C. neoformans* var. *neoformans*) do not. Thus, reproduction in *C. neoformans* is both possible by asexual budding, the production of basidiospores and the sexual cohesion of the two recognized mating types.

## 9.2 Ecology

*Cryptococcus neoformans* is a ubiquitous organism that is found in a variety of environments typically soil, decayed wood and avian excreta (e.g., pigeon droppings) or their nesting places such as window ledges and barns. Although *C. neoformans* grow at high densities in pigeon faeces, however birds are not infected by these pathogens. They are also found on fruits and other spoiled foods, damp surfaces and air. Of note, serotype A and D have been found throughout the world, while serotypes B and C are more frequently found in tropical and subtropical areas.

## 9.3 Diagnosis

The symptoms are the initial indicative of the CNS involvement and are confirmed with the CSF findings including elevated pressure, low glucose, increased protein concentration ad leukocytes counts of 20/mm$^3$ or higher. However, these CSF abnormalities may be absent in AIDS patients. The confirmatory clinical diagnosis is achieved by demonstration of the organism in the infected tissue, e.g., CSF, skin lesions or lung biopsies. The specimens are mixed with India ink or nigrosin and observed under a microscope. In addition, specimens should be cultured in niger seed agar for confirmatory diagnosis. Serological tests include the detection of capsule antigens. Latex agglutination detects antigens from the CSF and the serum from more than 90% of cryptococcal meningitis patients. In addition, ELISA tests are available for the rapid and sensitive detection of cryptococcal meningitis. Serology-based assays are targetted to detect cryptococcal antigens as the antibody levels may be undetectable.

## 9.4 Pathogenesis and Pathophysiology of *C. neoformans* CNS Infections

As described above, *C. neoformans* is a ubiquitous organism indicating our frequent exposure to these opportunistic pathogens. The term 'opportunistic' is referred to describe the need of predisposing factors for *C. neoformans* to produce infections such as a weak immune system. For example, *C. neoformans* rarely causes infection in individuals with CD4 T-cells count of more than 100 per µl. *Cryptococcus* infections although occur worldwide, are particularly abundant in regions with large populations of AIDS patients. *C. neoformans* usually enter the body through broken skin or the lungs. Skin infections are largely ignored but may develop into papules and cellulites in AIDS patients. The involvement of the lungs in AIDS patients may result in adult respiratory distress syndrome with acute pneumonia. Infection of the CNS is thought to initiate via inhalation of basidiospores. The *C. neoformans* yeast cell is approximately 5 µm, too large to pass through the alveoli into the primary site of infection, the parenchyma is in the lung, hence, it is suspected that the basidiospore, which is less than half the size, is the true culprit and gains access into the bloodstream. Alternatively skin lesions may provide a direct access route into the blood. It is not clear how *C. neoformans* survives in the bloodstream and it is presumed that polysaccharide capsule may play a role in cryptococcal persistence/survival in the bloodstream. Following invasion of the bloodstream, *C. neoformans* disseminates to other tissues including the liver, bone marrow, and the brain to produce a CNS infection. Again, cryptococcal traversal of the blood-brain barrier is not known and is the subject of future studies.

## 9.5 CNS Infections

The CNS infection due to *C. neoformans* is rarely associated with extensive hemorrhage, infarcion, calcification, extensive fibrosis or extensive tissue damage. The inflammatory response range from mild, moderate to strong. Macrophages outside the CNS (rarely in the meninges) are seen associated with cryptococci. *C. neoformans* form characteristic lesions throughout the brain with cortical grey matter most heavily involved but induce very limited inflammatory response. Since infection involves both meninges and the brain, it is more appropriately termed meningoencephalitis instead of meningitis. Occasionally, focal collection of *C. neoformans* and the inflammatory cells may occur in the brain, called cryptococcal granuloma but normally these infections are diffused. Patients exhibit symptoms such as headaches, nausea, dizziness, irritability, confusion, fever, pneumonia, vomiting, photophobia, seizures, coma and finally death. In addition, hydrocepalus may add complexity and result in increased intracranial pressure. In such cases, computer tomography (CT scan) or magnetic resonance imaging (MRI scan) are useful in confirming the involvement of the CNS. Of interest, *C. neoformans* var. *gattii* are more frequently associated with cerebral or pulmonary infections in non-HIV individuals and has low mortality. In contrast, *C. neoformans* var. *neoformans* are frequently associated with HIV patients and produce high mortality. The treatment includes the use of amphotericin B for few weeks followed by the application of fluconazole for several weeks. Despite the treatment, AIDS patients rarely survive this infection. Of the cured, up to 40% develop neurologic deficits including visual impairment, cranial nerve palsies, motor impairment and decreased mental function. Fluconazole may also be used as prophylactic agents in AIDS patients whose CD4 T-cell count is less than 50 cells per μl.

The capsule is a major virulence factor of *C. neoformans* infections. For example, capsular polysaccharide play a role in the resistance to phagocytosis by the immune cells. In addition, the capsule suppresses leukocyte migration and inhibits neutrophil attachment to the vascular endothelium by causing the shedding of L-selectin from the neutrophil surface. This inhibits neutrophil migration from the bloodstream into the infected tissues. In addition, the capsule inhibits T-cell responses and the antigen-presenting capacity of the macrophages. Although, capsular polysachharide induce complement activation but require large amounts of capsules, which is rarely found in infected individuals and only found in patients with severe cryptococcosis. Thus, the capsule may aid fungus, at least initially, in the evasion of the complement. The composition of the capsule is reminiscent of the cellular wall found in Gram-negative bacteria and is composed of two polymers. Approximately 90% of the polysaccharide is derived from glucuronoxylomannan (GXM), a negatively charged polymer knitted together by repetitive trimers of mannose. At intervals, mannose residues are 6-O-acetylated. A glucoronic-substituted backbone exhibits variation in mannose to xylose ratios between the recognized serotypes conferring consequential diversification in terms of antigenic response. The

GXM component of the polysaccharide capsule is a key virulent factor found to stimulate infectivity by decreasing pro-inflammatory levels of tumour necrosis factor-α (TNF) and interferon *r* by as much as 50%, whilst increasing anti-inflammatory cytokine production of IL-4 and IL-10. In contrast, the second polysaccharide GalXM is derived from galactose straddled by alternating mannose and xylose side chains, and occupies roughly 7% of the capsule mass. The remaining proportion of the envelope is procured from a mannoprotein. It is thought that the construction of the polysaccharide may employ as many as 11 enzymes and involve a UDP biosynthetic pathway. The capsular synthesis require a G-protein (GPA1) in concert with adenylate cyclase and cAMP to regulate the polysaccharide in *C. neoformans*, polysaccharide material being transported to the external cellular media via vesicles.

Other virulence factors of *Cryptococcus* include melanin since melanin mutants exhibit decreased virulence. Melanin is deposited in the inner cell wall and may contribute to the cell wall integrity and may act in providing resistance to oxidant produced during phagocytosis as well as decreased susceptibility to antifungal agents such as amphotericin B. In addition, *C. neoformans* produces mannitol, which has the potential for swelling of the brain (cerebral edema) as well as quenching the oxygen radicals, thus may also be important in the inhibition of phagocytic killing. The ability to grow at high temperatures (37°C) is an important parameter in the pathogenicity of *C. neoformans*. Recent studies have shown that CSF growth of *C. neoformans* requires the expression of ADE2 gene encoding phosphoribosylaminoimidazole carboxylase. Similarly the ability to produce phospholipases and proteases may have potential roles in the pathogenicity of *C. neoformans*. Future studies will precisely determine the role of each virulence determinant together with their mechanisms, which should help in the rationale development of therapeutic interventions. Other important factors include *C. neoformans* adaptive strategists, i.e., capable of reversible phenotypic switching. *C. neoformans* colonies switch between smooth, mucoid, wrinkled, serrated spores and hyphal structures. *In vitro* studies indicate that morphological changes in shape and size, transition from budding yeast to filamentous hyphae may contribute to virulence, reduced susceptibility to macrophage engulfment and dispersal and survival under harsh conditions. *C. neoformans* rely on cAMP to manipulate morphological transition and capsular regulation respectively.

## 9.6 Host Defence Mechanisms

The immune system of healthy individuals is highly effective in clearing cryptococci from the human body and the development of protective immunity against these pathogens. The primary circulating neutrophils as well as macrophages can effectively kill cryptococci by oxidative and non-oxidative mechanisms as shown by *in vitro* studies. In the process, cytokines are released that enhance the effectiveness of the anti-cryptococcal immune responses. To this end, TNF-$\alpha$, IL-12, IL-18, granulocyte-macrophage colony-stimulating factor, interferon-gamma, macrophage inhibitory protein-1alpha, macrophage chemotactic protein-1 and other cytokines have shown to be important in building an effective anti-cryptococcal immune response. However, more virulent *C. neoforman* strains may be able to evade this response or stimulate anti-inflammatory responses as indicated above. The humoral response also plays an important role in clearing cryptococcal infections. Antibodies opsonize cryptococci to enhance the phagocytic activity of the immune cells (antibody-dependent cell-mediated killing) and complement activation as well as clear the help clear the circulating antigens from the human body.

*C. neoformans* and *C. albicans* can equally survive at low or high temperatures, fluctuations prompting enhanced transcription of either histones or heat shock proteins. Consequently the transition from the external environment to an internal temperature of 37°C is not deleterious. Changes in temperature, osmotic pressure and pH will initiate a decrease or increase in the *C. neoformans* capsular size, chronic infection revealing both biochemical and structural alteration. The polysaccharide capsule carries a copper-dependent laccase and secretes a phospholipase that may mediate during virulence by degrading the phospholipid composition of the host membrane, though secretion of other enzymes has similarly been highlighted, such as urease and proteinases. The enzyme laccase is of particular interest as it catalyses the production of melanin providing an exogenous substrate is available. Therefore if the pigment is to provide cell wall protection synthesis must be performed by substrates similar to *O*-diphenol where the hydroxyl groups occupy positions 2, 3- or 3,4, so enabling hydrolytic enzymes (fungal) to be secured within the parameters of the cell wall. Compared to their non-melanised counterpart Mel$^+$ *C. neoformans* exhibit anti-oxidant resistance, reduced damage by heavy metals, specifically iron, and a decreased susceptibility to ingestion by alveolar macrophages. Further, pigmentation reduces environmental damage by ultra-violet light and notably impedes the clinical effects of the anti-fungal drug amphotericin B. Wild-type melanin forms of *C. neoformans* are infinitely more virulent. Indeed Mel+ *C. neoformans* have been observed in the brain tissue, but are equally common to pigeon excreta, suggesting that melanization may precede infection in some cases.

With increased virulence *C. neoformans* produces mannitol providing added protection against dehydration and degradation from phagocytosis or neutrophilic attacks. Raised levels of mannitol in the CNS may well contribute to virulence by encouraging brain oedema. Brain tissue and the area surrounding the basal ganglia is an environment for numerous catecholamines or neurotransmitters, such as dopamine, noradrenaline and norepinephrine. The laccase converts these catecholamines to melanin and in so doing prevents toxic damage occurring to the fungus, presumably either at the outset or once infection has taken hold, though it could well occur during a latent, incubatory phase. Further, the laccase participates in a FeII to FeIII conversion that has been demonstrated during *in vitro* studies to protect *C. neoformans* from attack by hydroxyl radicals. Ironically, PCA-2 a low virulent strain of *C. albicans* has been used to stimulate the production of astrocytes and microglial cells in the brain to mount an anticryptococcal defensive, though success relies on the offender (*C. neoformans*) being under antibody attack and therefore opsonized. Since melanins are negatively charged hydrophobic molecules and the cell wall of *C. neoformans* includes repetitive chains of galactose linked by terminal sialic acids (N-acetylneuraminic acid), the overall negative charge is dramatically increased, thereby enhancing electrostatic forces vital to cell adhesion and contributing to inhibition of galactose receptors on circulating macrophages.

**Table 1** Properties of fungi.

| *Fungi* | *Asexual structure* | *Sexual spore* | *Example* |
|---|---|---|---|
| Chytridiomycota | Spores, conidia | Oospore | Allomyces spp. |
| Zygomycota | Spores, conidia | Zygospore | Rhizopus |
| Ascomycetes | Conidia, blastoconidia | Ascospore | Saccharomyces |
| Badidiomycetes | Basidiospores | Basidiospores | Amantia (poisonous mushroom)<br>Agaricus (edible mushroom) |
| Chromista<br>(Fungi imperfecti) | Conidia, blastoconidia<br>phialoconidia | None | Penicillium, Aspergillus, Candida |

CHAPTER

# 7

# Microbes as Bioweapons

## 1. MICROBES AS BIOLOGICAL WEAPONS

The use of germs to kill civilians is bioterrorism and their use to kill enemy forces is biowarfare. It has become clear that bioweapons are a potential threat to human kind. The most worrisome aspect is that unlike nuclear or chemical weapons, military arsenal, high-tech machinery which are normally state-sponsored, bioweapons can be used with minimal resources by states or even by individuals.

## 2. HISTORY

The use of germs as warfare agents is nothing novel. This method of killing enemy forces or civilians has been used throughout our history. For example, in 184 BC, Hannibal ordered serpent-filled pots to be thrown onto the decks of enemy ships. In the 1340s, the Tattar army used plague-infected people to spread infection to aid in their conquest. In the 1760s, the British army provided the American Indians smallpox infected blankets. More recently, in World War I, Germany used pathogens as agents of germ warfare in Europe. Following World War I, the biggest players in bioweapons were the USA, Japan, USSR, Germany, UK and Canada but their reported use is unclear. It is well-established that in World War II, Japan used bioweapons (plague, typhoid, cholera, anthrax and other agents) against the Chinese and the Soviets. During World War II, England conducted experiments on the use of bioweapons (*Bacillus anthracis*) on Gruinard Island (near the coast of Scotland). This resulted in heavy contamination of the Island, which only has recently (1986) been decontaminated with seawater and formaldehyde. Iraq's pursuits of bioweapons lasted from 1975 to 2003, while South Africa had a limited programme for bioweapons from 1981 to 1993. At present, it is unclear how many nations (probably around a dozen) are actively seeking the most sophisticated bioweapons but what is clear is that bioweapons present a clear and present danger.

## 3. AGENTS FOR BIOWEAPONS

Despite the clear devastating effects of bioweapons on humans, animals, plants and the environment, some continued to build biological weapons or pursued research to construct the most lethal microbes/toxins. As the danger of their use is real, there is a clear need to identify microbial agents/toxins that can be used as biological weapons, their mode of action and be prepared in the unfortunate event of a bioterror attack. Potentially any microbe or microbe-derived toxin capable of causing harm or human misery is a bioweapon agent. Among others, smallpox virus, anthrax and botulism toxins are considered the most significant bioweapons. To this end, the Centers for Disease Control and Prevention (CDC) in Atlanta, GA, USA published a list of critical biological agents in 2000 and divided them into three categories (Fig. 1). The most likely entry of these pathogens/toxins is inhalation into the lungs, ingestion of contaminated food or water or absorption of toxins through the skin presenting an easy mode of transmission to a large number of people. Of interest, the category A agents are most effectively transmitted via aerosols and are the most devastating bioweapons. In addition, our ability to genetically modify organisms may present serious consequences. For example, it is shown that genetic modification of mousepox virus can have fatal consequences for mice, which are normally resistant to this virus. Among category A, anthrax has gained the most attention in recent years. This is due to anthrax (*Bacillus anthracis*) spores enclosed in a letter and mailed to the US media and government officials in 2001 which resulted in several fatalities. The outcome confirmed the dangers of bioweapons but more importantly the simplicity in their use and dissemination. The potential of anthrax as a bioweapon is due to its ability to readily form spores which can be stored indefinitely and can survive in the dry air for several hours. They are Gram-positive bacteria and can cause serious gut (gastrointestinal anthrax) or skin infections (cutaneous anthrax) but if inhaled, they can produce inhalational anthrax with fatal consequences (mortality rate of more than 90%). The incubation period varies from a few hours to a few days. The clinical symptoms include fever, malaise, fatigue, respiratory failure, septicemia and finally death within 24 h (even with treatment). As anthrax is non-contagious in humans, their effects are limited to the applied area. In contrast, the smallpox virus is highly contagious and can result in global pandemics with a mortality risk of 30%. *Bacillus anthracis* is normally present in soil worldwide, and there are approx. 200 – 2,000 annual cases of anthrax (mostly cutaneous). It is the inhalation of anthrax that is most associated with bioterror and has the potential of a bioweapon.

## 4. BIODEFENCE

Politicans aided by scientists developed bioweapons. Similarly, in defending civilization, politicians with the help of scientists must actively and urgently devise plans for eliminating this potential threat. The approaches to defend against bioweapons can be divided into two groups.

## 5. PREVENTATIVE MEASURES

These include international cooperation, monitoring all states for biological warfare programmes, enforcement of non-proliferation of bioweapons treaties for all nations (including Superpowers) and the availability of vaccines against potential bioweapon agents. These approaches urgently require international diplomacy and cooperation, proactive role of independent monitoring agencies and establishments of international enforcement agencies. As with other warfare, the justification of biological warfare programmes is bizarre. For example, the only purpose a bullet will serve is to kill, whether a friend or foe. The justification of making a bullet is in the eyes of the beholder, whether defence or offence. What we all must understand is that bioweapons is a common threat to human kind and it need not be added on the extensive list of weapons at our disposal to kill our fellow human beings for narrow political gains. Other more applicable preventative approaches include vaccines. Vaccines are useful tools against bioweapons and entire populations must be vaccinated but they have no value as therapeutic measures. In addition, vaccines may have potential side effects and may require many doses. For example, the present anthrax vaccine requires six shots over a period of two years. Also, the genetic manipulation of the infectious agents or use of different strains of agents may render the vaccine obsolete.

## 6. THERAPEUTIC MEASURES

The success of therapeutic approaches relies on the availability of rapid diagnostic tests. Accurate early detection followed by appropriate treatment including the use of antibiotics, antiviral, and antibodies for passive protection may have protective effects. Antibiotics and antivirals have been the most effective mode of treating infectious agents. But again, with the advances in genetic approaches, the development of antimicrobial-resistant

Category A
- Easily disseminated or transmitted person-to-person
- Cause high mortality
- Require special action for public health preparedness

Bacteria

- *Bacillus anthracis* (anthrax)
- *Yersinia pestis* (plague)
- *Francisella tularensis* (tularemia)
- *Clostridium botulinum* toxin (botulism)

Viruses

- Variola major virus (smallpox)
- Viruses causing haemorrhagic fevers such as Ebola, Marburg, Lassa, Hanta, Congo, South American, Rift valley, Tick-borne and Yellow fever

Category B
- Moderately easy to disseminate
- Cause moderate morbidity and low mortality
- Require specific enhancements of diagnostic capacity and disease surveillance

Bacteria

- *Coxiella burnetti* (Q fever)
- *Brucella* species (brucellosis)
- *Burkholderia mallei* (glanders)
- Epsilon toxin of *Clostridium perfringens*
- *Staphylococcus* enterotoxin B
- Foodborne or waterborne agents also are included such as pathogenic *Salmonella* spp. *Shigella* spp. *Escherichia coli* spp. *Vibrio cholerae*

Viruses

- Alphaviruses (Venezuelan encephalomyelitis and eastern and western equine encephalomyelitis)

Category C
- Pathogens that could be engineered for mass dissemination in the future because of: availability, ease of production and dissemination
- Potential for high morbidity and mortality and major health impact

Bacteria

- Multidrug-resistant *Mycobacterium tuberculosis*

Viruses
- Nipah virus
- Hantaviruses
- Tickborne haemorrhagic fever viruses
- Tickborne encephalitis viruses, Yellow fever virus

**Fig. 1** List of biological agents that can be used as bioweapons (source www.cdc.gov).

strains is relatively simple and an unfortunate opportunity to construct more deadly pathogens. Passive protection using antibodies is an effective therapeutic approach, especially in neutralizing toxins such as botulism toxin, ricin (made from waste of processed castor beans), epsilon toxin of *Clostridium perfringens*, Staphylococcal enterotoxin B. However, the use of different strains is a potential hurdle in our preparedness against bioweapons. Any bioterror attack will have serious implications on the environment. The need to develop measures for environmental cleanup is urgent. At present, we have limited if any environmentally friendly disinfectants or measures to assess the tolerable levels of contamination or infrastructure to perform such massive tasks. Future work must involve international diplomacy and cooperation, the enforcement of non-proliferation of bioweapons treaties for all nations, identification of novel antimicrobials, research in the understanding of the pathogenesis and pathophysiology of diseases caused by bioweapons, environmental assessment and cleanup. These measures may aid in our ability to combat microbial monsters created for our narrow political gains.

# Index

## N



## O



## P



## Q



## R

Chapter 2

# Colour Plate Section

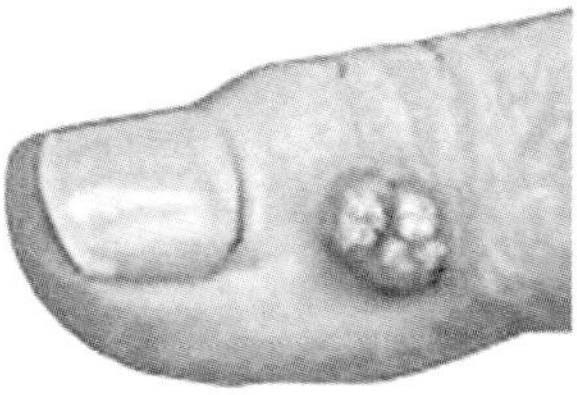

**Fig. 3** Cutaneous wart (source: www.nlm.nih.gov/medlineplus).

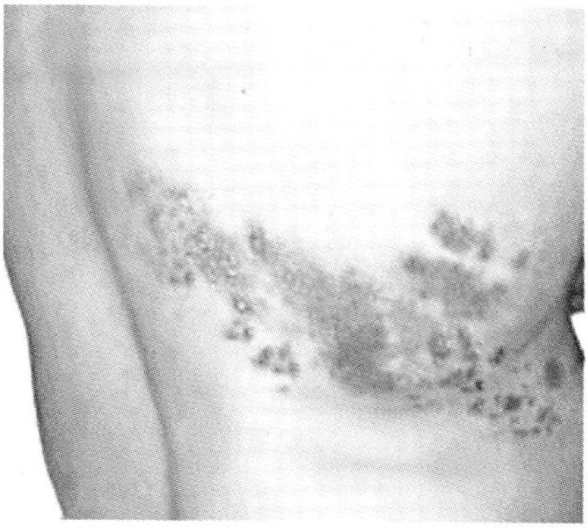

**Fig. 4** Shingles (herpes zoster) (source: www.nlm.nih.gov/medlineplus).

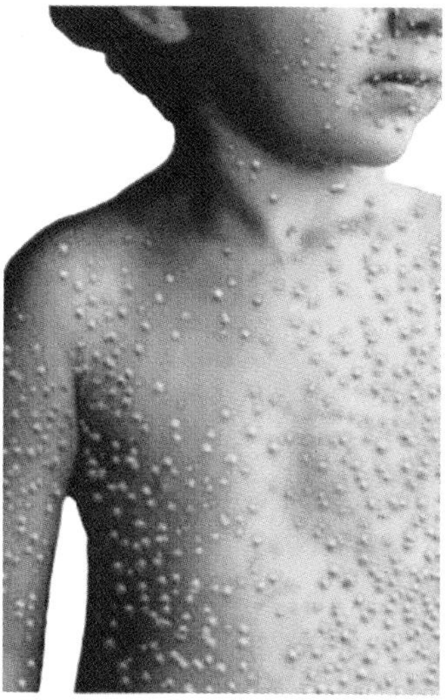

**Fig. 5** Smallpox-infected patient (source: www.nlm.nih.gov/medlineplus).

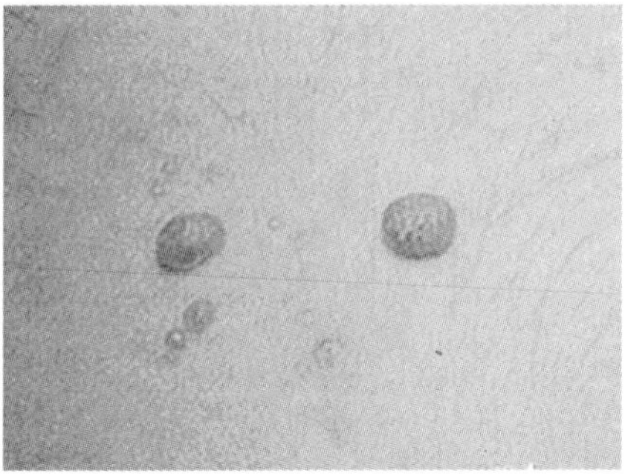

**Fig. 6** Molluscum contagiosum infection (source: www.nlm.nih.gov/medlineplus).

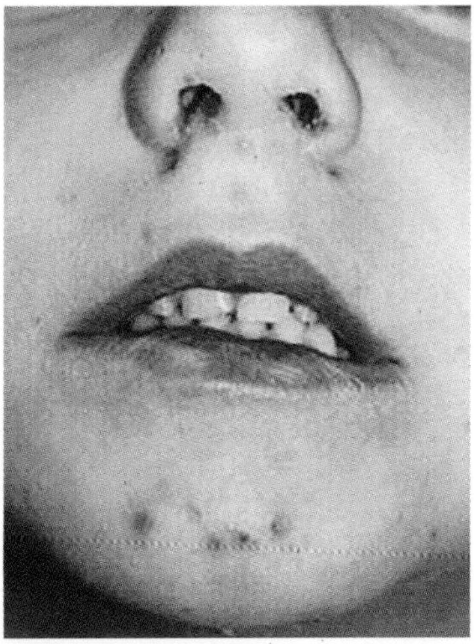

**Fig. 7** Impetigo-infected child (source: www.nlm.nih.gov/medlineplus).

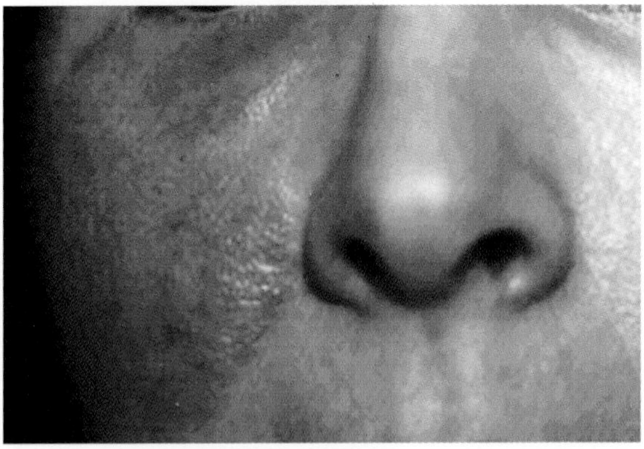

**Fig. 8** Erysipelas--infected patient (source: www.nlm.nih.gov/medlineplus).

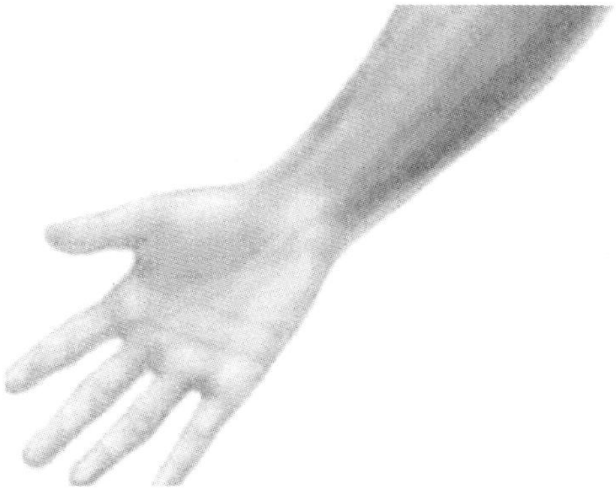

**Fig. 9** Skin lesions due to cellulitis (source: www.nlm.nih.gov/medlineplus).

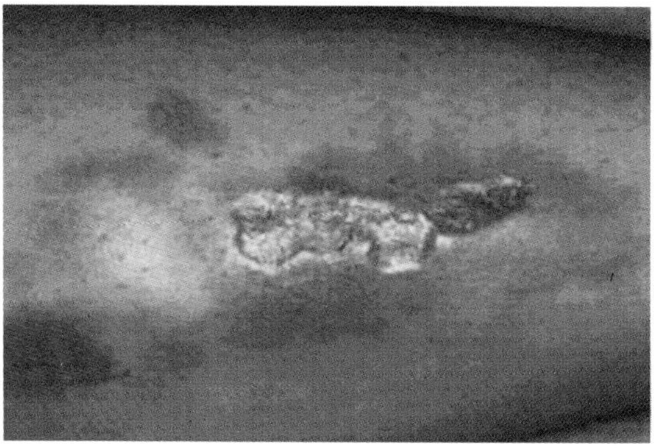

**Fig. 10** Skin lesions due to necrotizing fasciitis (source: www.nlm.nih.gov/medlineplus).

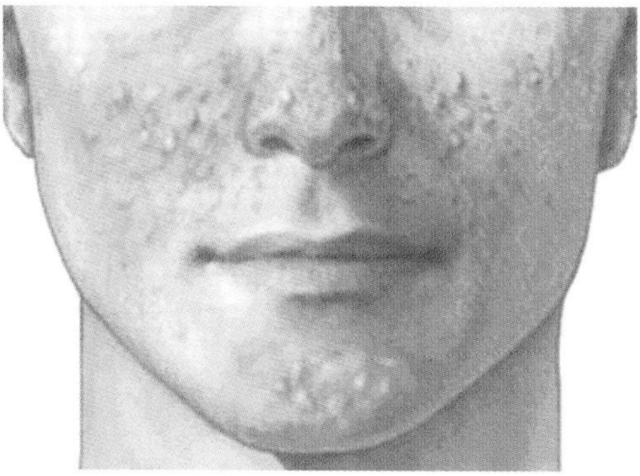

**Fig. 11** Acne, a common skin infection (source: www.nlm.nih.gov/medlineplus).

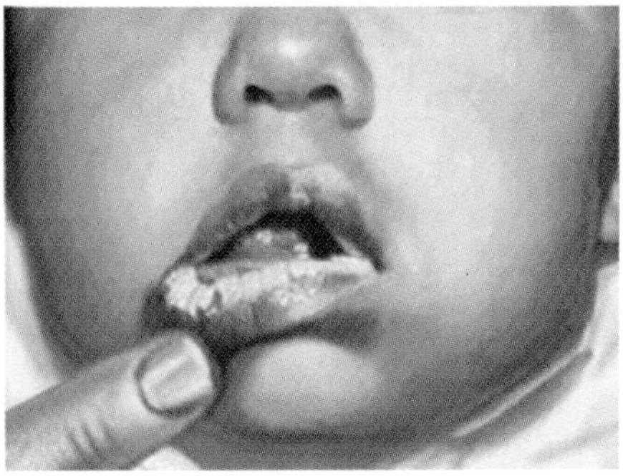

**Fig. 12** Candidiasis of the skin (source: www.nlm.nih.gov/medlineplus).

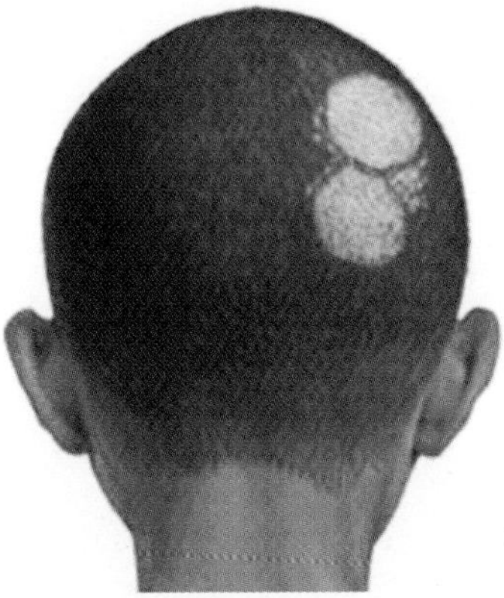

**Fig. 13** Tinea infection (source: www.nlm.nih.gov/medlineplus).

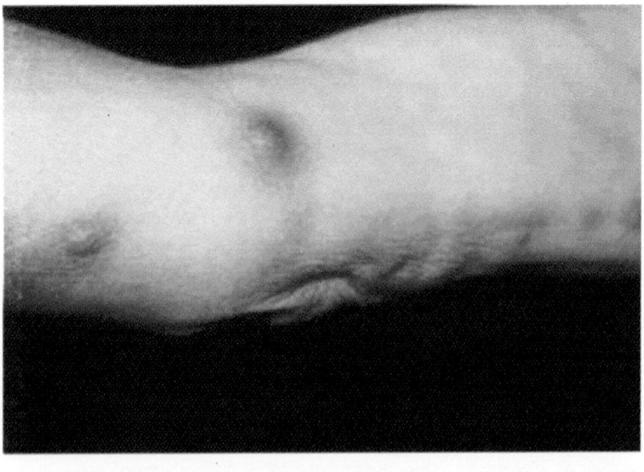

**Fig. 14** Sporotrichosis-infected patient (source: www.nlm.nih.gov/medlineplus).

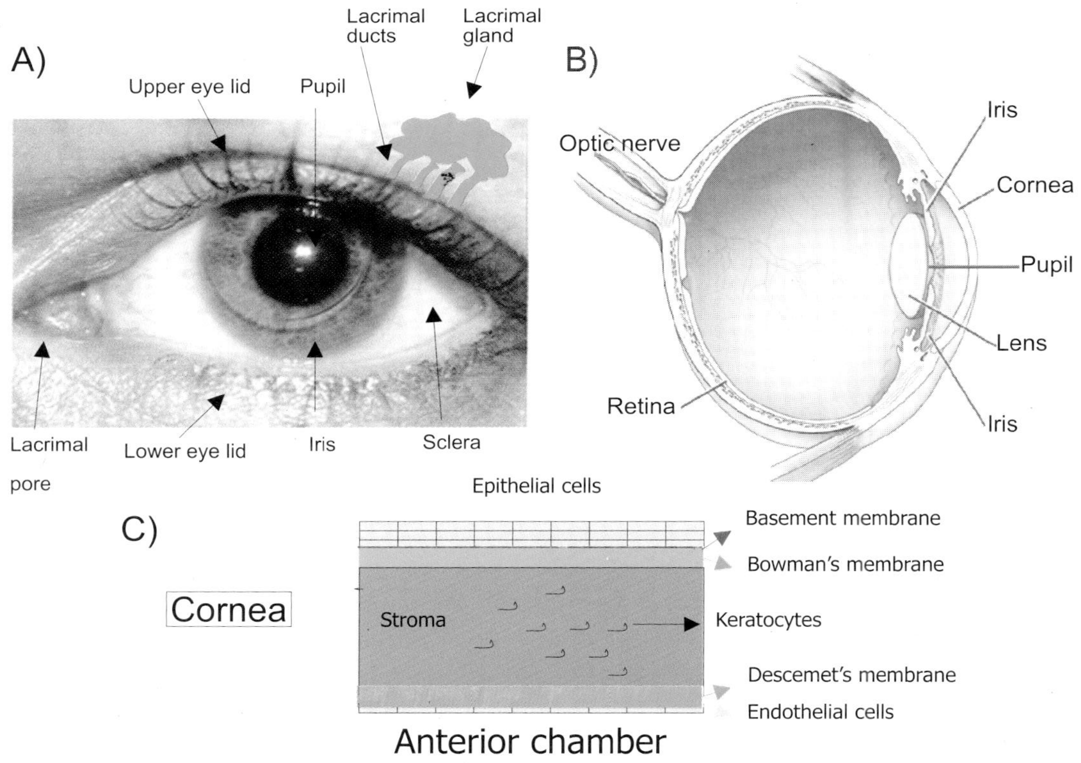


**Fig. 15** (A and B) Anatomy of the eye and (C) cellular structure of cornea.

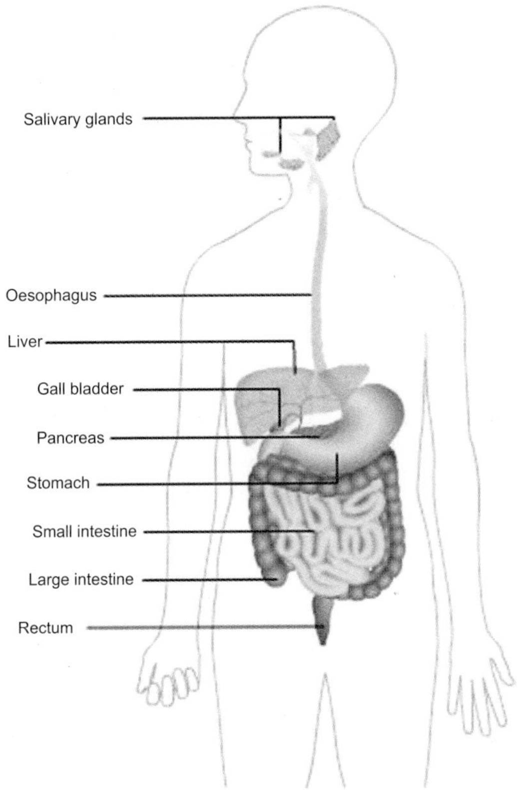


**Fig. 18** Gastrointestinal system.

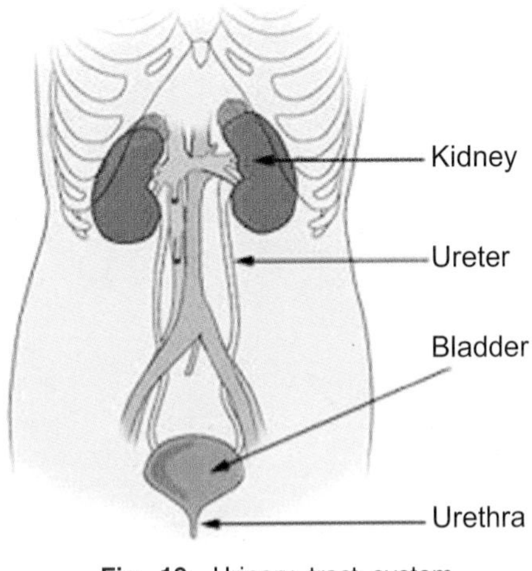


**Fig. 19** Urinary tract system.

(A)

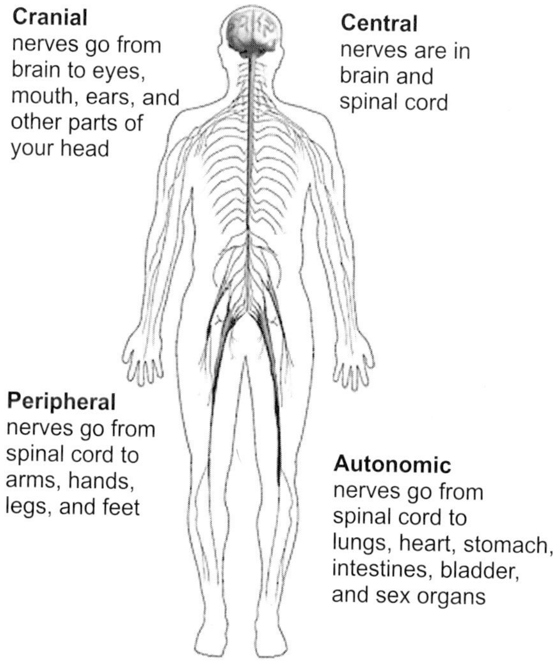
**Cranial**
nerves go from brain to eyes, mouth, ears, and other parts of your head
**Central**
nerves are in brain and spinal cord
**Peripheral**
nerves go from spinal cord to arms, hands, legs, and feet
**Autonomic**
nerves go from spinal cord to lungs, heart, stomach, intestines, bladder, and sex organs

B)

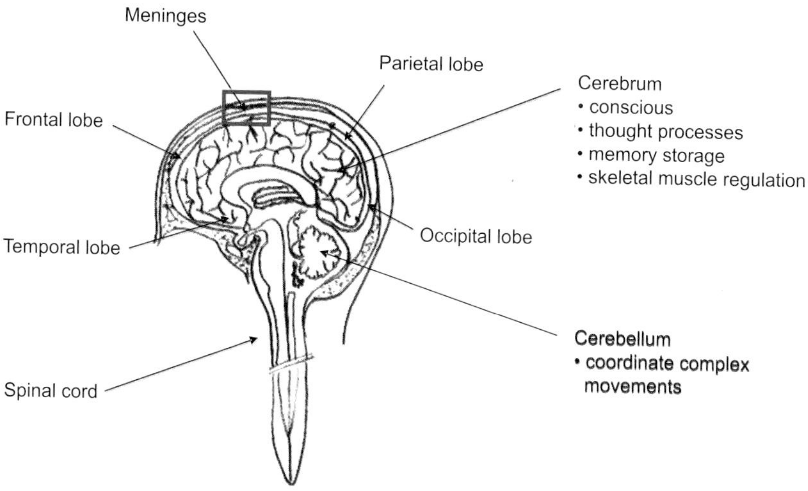
Meninges
Parietal lobe
Cerebrum
• conscious
• thought processes
• memory storage
• skeletal muscle regulation
Frontal lobe
Temporal lobe
Occipital lobe
Cerebellum
• coordinate complex movements
Spinal cord

(C)

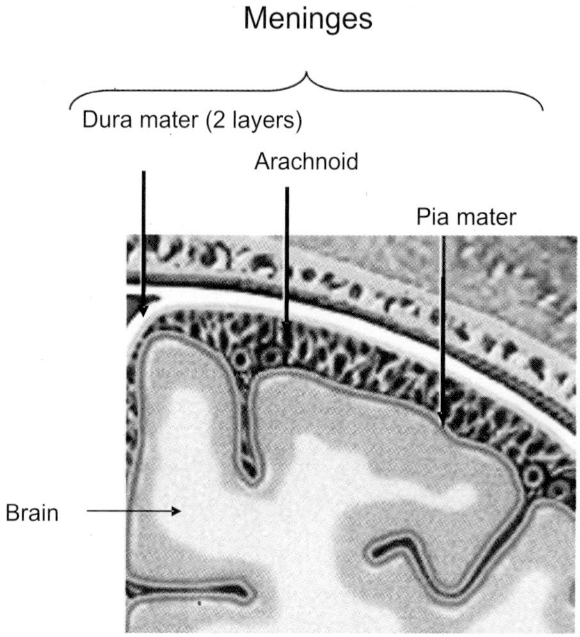


**Fig. 20** (A) Nervous system, (B) Structure of the central nervous system, and (C) meninges, i.e., membranes covering the brain and the spinal cord.

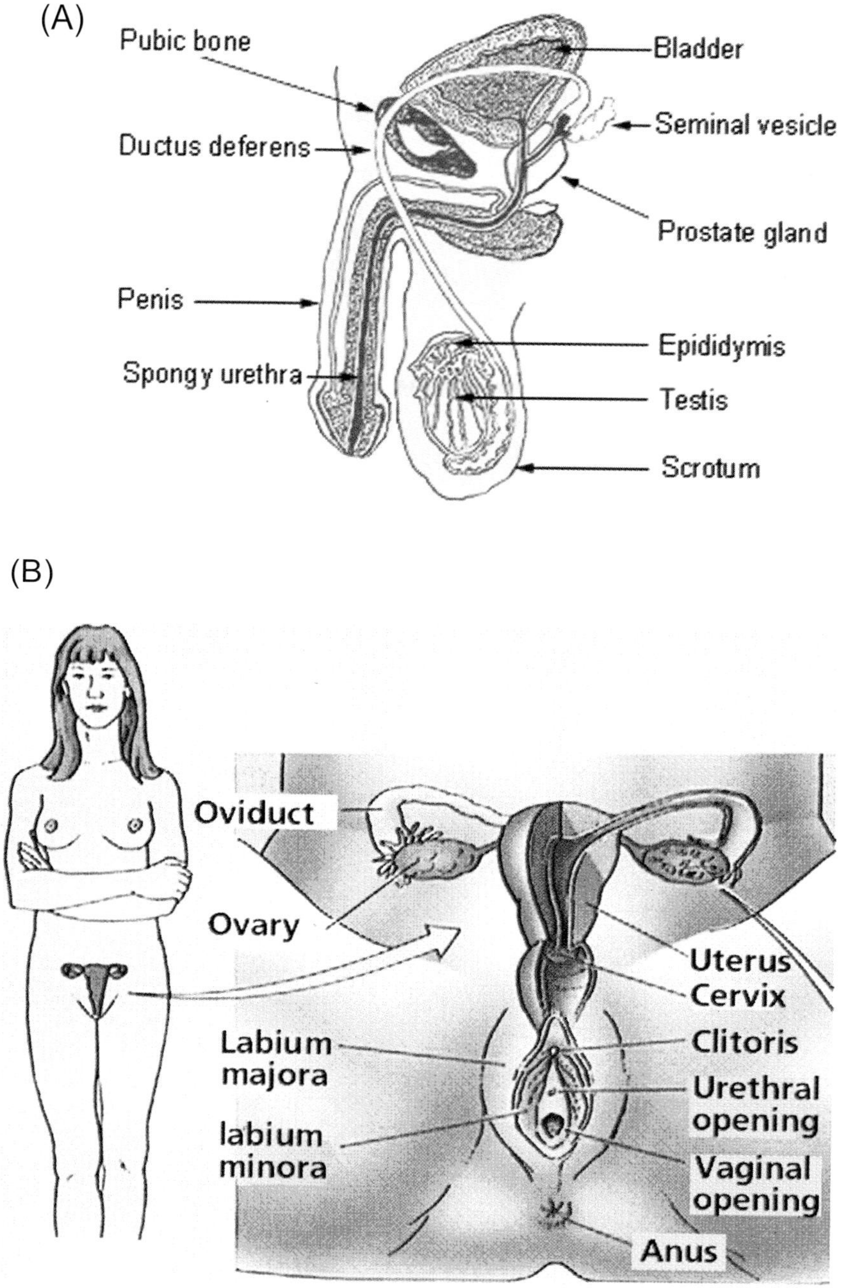


**Fig. 21** Human reproductive system, (A) male and (B) female.

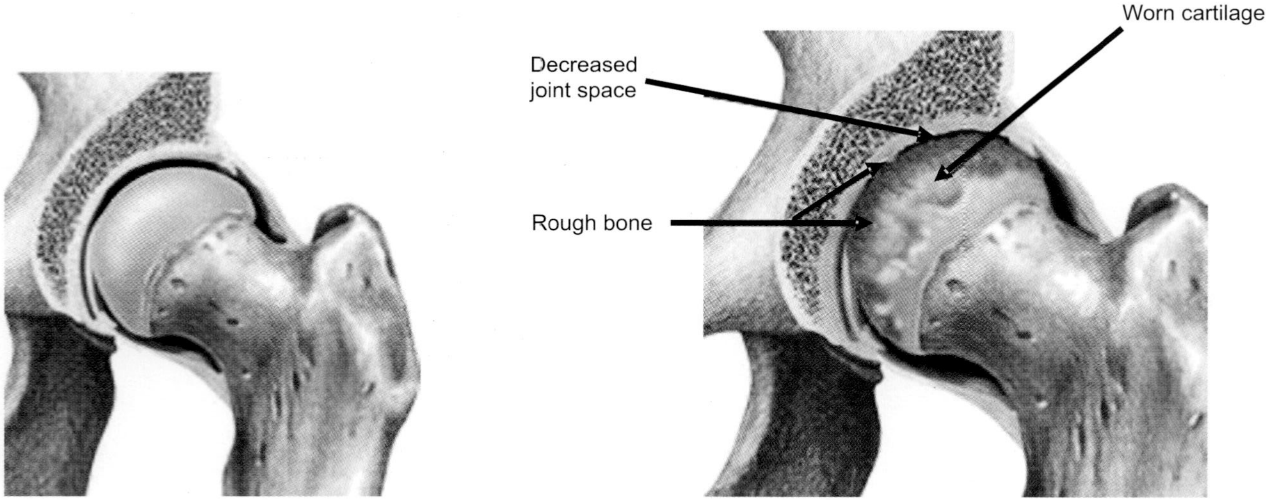


Normal hip joint Arthritic hip joint

**Fig. 23** Breakdown of cartilage in arthritis (source: www.nlm.nih.gov/medlineplus).

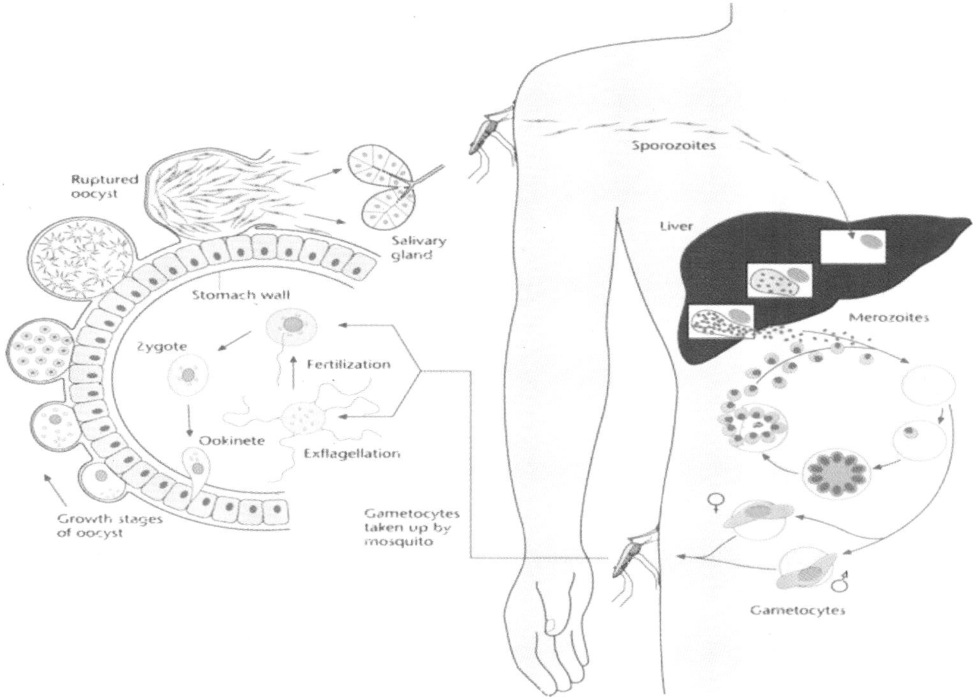


## Chapter 5

**Fig. 4** The life cycle of *Plasmodium* spp. indicating vertebrate and invertebrate hosts.

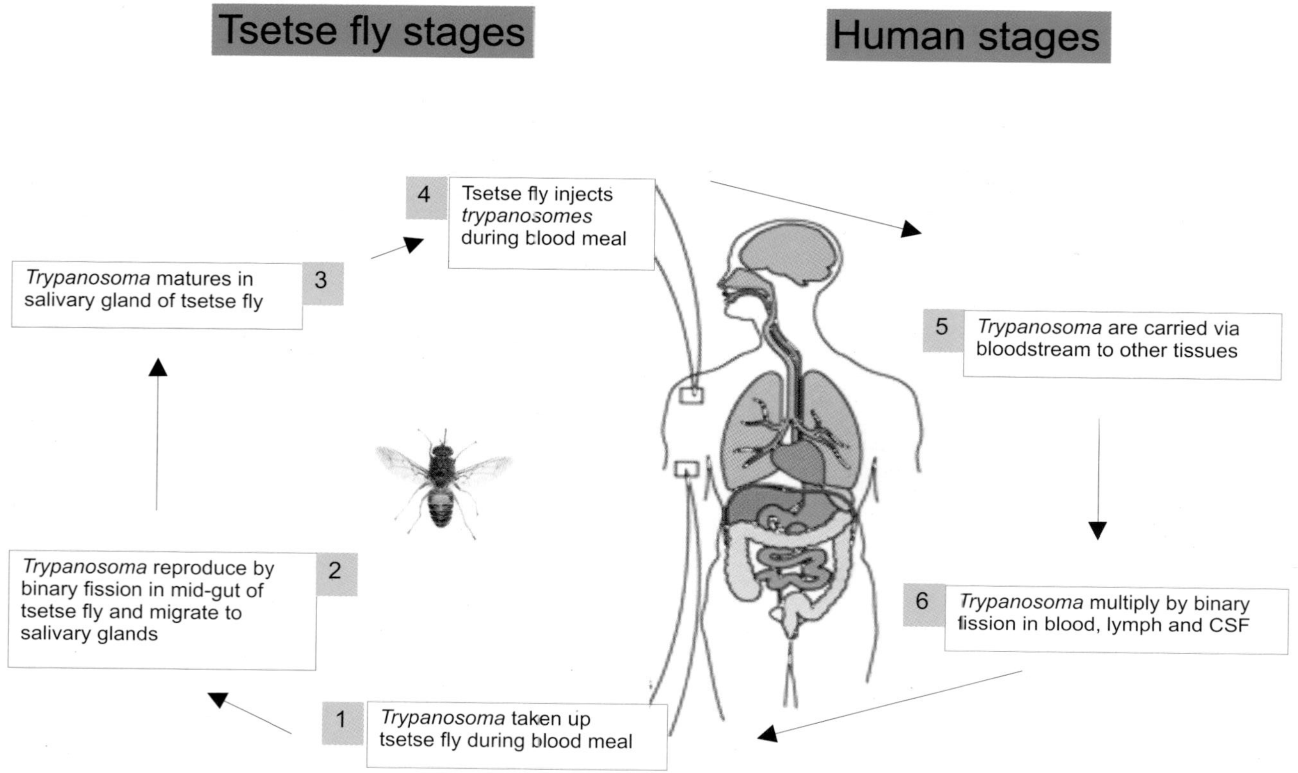


**Fig. 5** The life cycle of *Trypanosoma brucei* indicating vertebrate and invertebrate hosts.

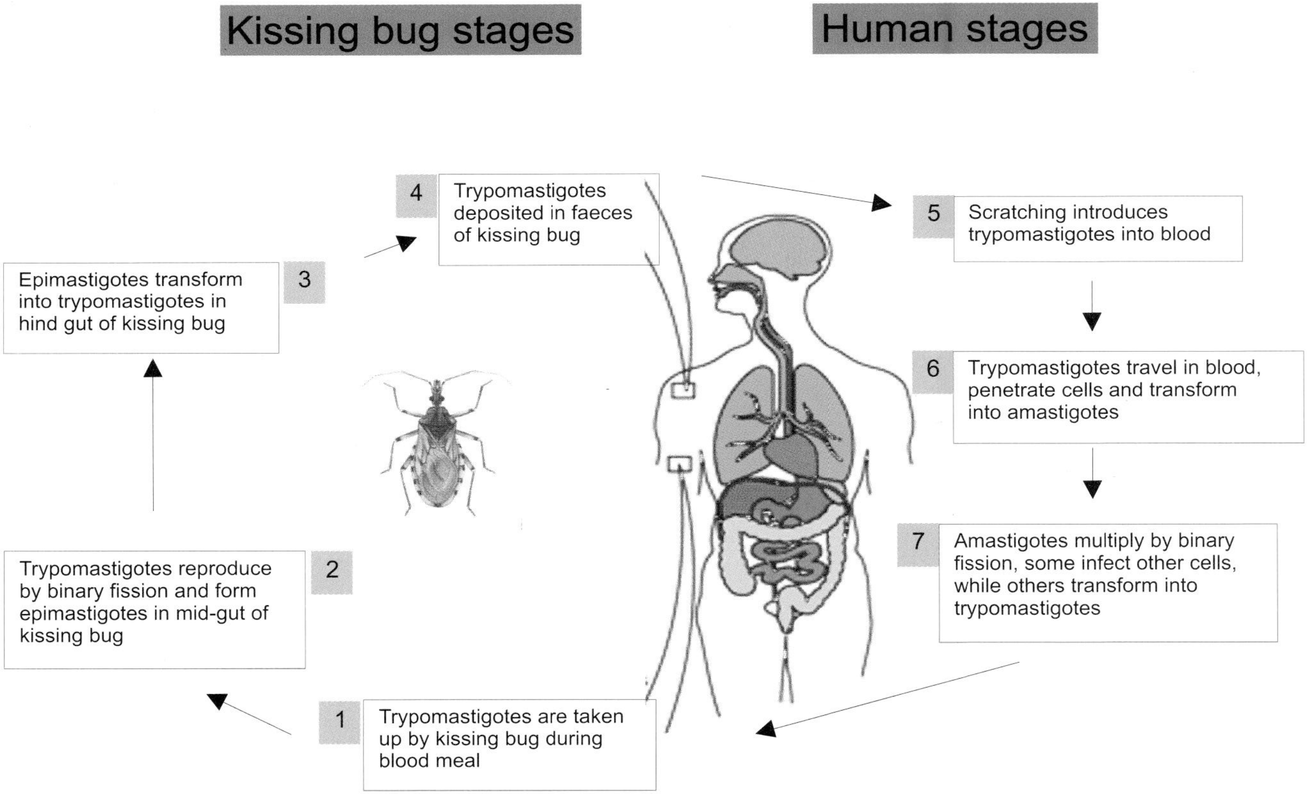


**Fig. 6** The life cycle of *Trypanosoma cruzi* indicating vertebrate and invertebrate hosts.

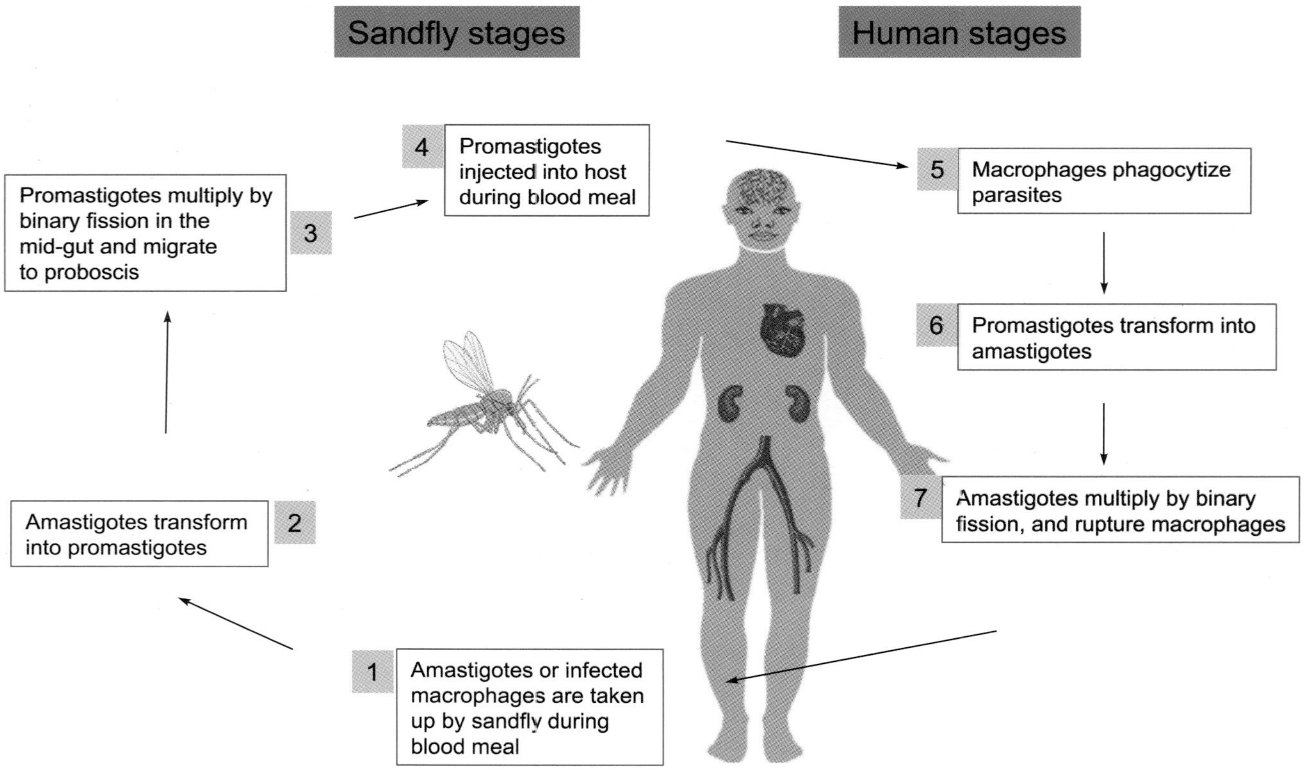


**Fig. 7** The life cycle of *Leishmania* spp. indicating vertebrate and invertebrate hosts.